Andreas Baierl

Model Selection Techniques for Locating Quantitative Trait Loci

Andreas Baierl

Model Selection Techniques for Locating Quantitative Trait Loci

Modifying the Bayesian Information Criterion for Locating Multiple Quantitative Trait Loci in Experimental Crosses

Südwestdeutscher Verlag für Hochschulschriften

Impressum/Imprint (nur für Deutschland/ only for Germany)
Bibliografische Information der Deutschen Nationalbibliothek: Die Deutsche Nationalbibliothek verzeichnet diese Publikation in der Deutschen Nationalbibliografie; detaillierte bibliografische Daten sind im Internet über http://dnb.d-nb.de abrufbar.
Alle in diesem Buch genannten Marken und Produktnamen unterliegen warenzeichen-, marken- oder patentrechtlichem Schutz bzw. sind Warenzeichen oder eingetragene Warenzeichen der jeweiligen Inhaber. Die Wiedergabe von Marken, Produktnamen, Gebrauchsnamen, Handelsnamen, Warenbezeichnungen u.s.w. in diesem Werk berechtigt auch ohne besondere Kennzeichnung nicht zu der Annahme, dass solche Namen im Sinne der Warenzeichen- und Markenschutzgesetzgebung als frei zu betrachten wären und daher von jedermann benutzt werden dürften.

Verlag: Südwestdeutscher Verlag für Hochschulschriften Aktiengesellschaft & Co. KG
Dudweiler Landstr. 99, 66123 Saarbrücken, Deutschland
Telefon +49 681 37 20 271-1, Telefax +49 681 37 20 271-0, Email: info@svh-verlag.de
Zugl.: Wien, Universität Wien, Diss., 2007

Herstellung in Deutschland:
Schaltungsdienst Lange o.H.G., Zehrensdorfer Str. 11, D-12277 Berlin
Books on Demand GmbH, Gutenbergring 53, D-22848 Norderstedt
Reha GmbH, Dudweiler Landstr. 99, D- 66123 Saarbrücken
ISBN: 978-3-8381-0122-4

Imprint (only for USA, GB)
Bibliographic information published by the Deutsche Nationalbibliothek: The Deutsche Nationalbibliothek lists this publication in the Deutsche Nationalbibliografie; detailed bibliographic data are available in the Internet at http://dnb.d-nb.de.
Any brand names and product names mentioned in this book are subject to trademark, brand or patent protection and are trademarks or registered trademarks of their respective holders. The use of brand names, product names, common names, trade names, product descriptions etc. even without
a particular marking in this works is in no way to be construed to mean that such names may be regarded as unrestricted in respect of trademark and brand protection legislation and could thus be used by anyone.

Publisher:
Südwestdeutscher Verlag für Hochschulschriften Aktiengesellschaft & Co. KG
Dudweiler Landstr. 99, 66123 Saarbrücken, Germany
Phone +49 681 37 20 271-1, Fax +49 681 37 20 271-0, Email: info@svh-verlag.de

Copyright © 2008 Südwestdeutscher Verlag für Hochschulschriften Aktiengesellschaft & Co. KG and licensors
All rights reserved. Saarbrücken 2008

Produced in USA and UK by:
Lightning Source Inc., 1246 Heil Quaker Blvd., La Vergne, TN 37086, USA
Lightning Source UK Ltd., Chapter House, Pitfield, Kiln Farm, Milton Keynes, MK11 3LW, GB
BookSurge, 7290 B. Investment Drive, North Charleston, SC 29418, USA
ISBN: 978-3-8381-0122-4

Preface

Statistical developments have in many cases been driven by applications in science. While genetics was always an important area that encouraged statistical research, recent technological advances in this discipline pose ever new and challenging problems to statisticians.

This thesis covers the application of model selection to QTL mapping, i.e. locating genes that influence a quantitative character of an organism. The existing statistical framework was taken as a starting point that has been adapted and modified to be applicable to QTL mapping. The requirements included good computational performance, robustness and consideration of prior information.

The key results from my thesis are covered in three joint publications:

- Baierl, A., Bogdan, M., Frommlet, F. and Futschik, A. (2006). On Locating Multiple Interacting Quantitative Trait Loci in Intercross Designs. *Genetics* **173**, 1693–1703

- Baierl, A., Futschik, A., Bogdan, M. and Biecek, P. (2007).Locating multiple interacting quantitative trait loci using robust model selection. *Computational Statistics and Data Analysis* **51**, 6423-6434

- Zak, M., Baierl, A., Bogdan, M. and Futschik, A. (2007). Locating multiple interacting quantitative trait loci using rank-based model selection. *Genetics* **176**, 1845–1854

Acknowledgements

I am grateful to my supervisor Andreas Futschik for introducing me to this fascinating area of research as well as for his constant guidance and stimulating advice. Andreas' statistical expertise and intuition as well as his focus on results had a great impact on the outcome of this thesis. Further, I want to thank Malgorzata Bogdan for her enthusiastic participation in our joint work and for sharing her broad knowledge on the topic.

Contents

1 **An Introduction To ...**   5
    1.1 Quantitative Genetics 5
        1.1.1 Quantitative Traits and Quantitative Trait Loci 6
        1.1.2 Historical Developments 7
    1.2 Model Selection 8
        1.2.1 Concept 8
        1.2.2 Model selection criteria 9
        1.2.3 Model selection vs. Hypothesis Testing 12
    1.3 Mapping of Quantitative Trait Loci 13
        1.3.1 QTL Mapping Techniques for Experimental Crosses 15
        1.3.2 QTL Mapping by Model Selection 17

2 **Locating Multiple Interacting Quantitative Trait Loci in Intercross Designs**   21
    2.1 Summary 22
    2.2 Introduction 22
    2.3 Methods 24
        2.3.1 Statistical Model 24
    2.4 A Modified BIC for Intercross Designs 25
    2.5 Simulations 30

2.6	Illustrations	41
2.7	Discussion	43
2.8	Appendix	46

3 Locating multiple interacting quantitative trait loci using robust model selection 48

3.1	Introduction	49
3.2	The statistical model	50
3.3	Robust model selection and the modified BIC	52
3.4	Comparison of performance under different error models	57
	3.4.1 Design of the simulations	57
	3.4.2 Error distributions	59
	3.4.3 Results of the simulations and discussion	59
3.5	Application to real data	64
3.6	Conclusions	66

4 Locating multiple interacting quantitative trait loci using rank-based model selection 67

4.1	Introduction	68
4.2	Methods	68
	4.2.1 Simulation design	69
	4.2.2 Simulation results	71
	4.2.3 Application to Real Data	75

Chapter 1

An Introduction To ...

1.1 Quantitative Genetics

Quantitative genetics, a statistical branch of genetics, tries to give a mechanistic understanding of the evolutionary process based upon the fundamental Mendelian principles. The goal of the evolutionary process, the optimal value of a trait, can be predicted nearly solely by natural selection. Issues that arise when trying to explain how the optimum is obtained, like

- the time it takes an optimal trait value to evolve
- how the genetic variation necessary for adaption arises
- the amount of expected phenotypic variation
- the role of non-adaptive evolutionary change caused by fluctuation of gene frequencies and mutations

are addressed by the discipline of quantitative genetics.

1.1.1 Quantitative Traits and Quantitative Trait Loci

In quantitative genetics, we study biological traits of individuals that are continuous or quantitative like the yield from an agricultural crop, the survival time of mice following an infection or the length of a pulse train, a song character, of fruit-flies.

The observed trait value (phenotype) P of an individual can be divided into the combined effect of all genetic effects G and and environment-dependent factors E:

$$P = G + E$$

The genetic effect can be based upon very few genes or even a single gene or it can be composed of a number of genes. Clearly, this distinction largely determines the distributions of the phenotype and has obvious consequences for the statistical methodology to be used. In fact, Mendel in his historic experiments on peas investigated traits like color or shape that were influenced by single or very few genes and were therefore qualitative. In quantitative genetics, we study traits that are influenced by several genes, i.e. are polygenic, and we call the locations of these genes quantitative trait loci (QTL). More precisely, a quantitative trait locus is not necessarily identical with the according gene, but can be any stretch of DNA in close distance to the gene. For polygenic traits, the phenotype can be either continuous or also discrete (e.g. counts, categories) in case of an underlying quantitative character with multiple threshold values.

In the case of polygenic traits, we divide the genetic effects further into separate contributions of single genes (additive and dominance effects) and the interaction between genes (epistasis). Obviously, the different genetic components cannot be estimated from a single observation, but by assessing a sample of individuals. In order to be able to separate genetic and environmental effects, the relatedness of the individuals has to be known. Here, an important distinction concerning the origin of the data has to be done: data from

- controlled programs imposed on domesticated species that are usually performed with specific sets of relatives of specific ages in specific environmental backgrounds

- natural populations like mammals, where we have a lack of experimental control.

1.1.2 Historical Developments

Francis Galton, a half-cousin of Darwin, founded the biometrical school of heredity by focusing on the evolution of continuously changing characters (Galton (1889, 1869)). The main principles of quantitative genetics have been outlined by R. A. Fisher (1918) and S. G. Wright (1921a,b,c,d). Both showed that the Mendelian principles can be generalized to quantitative traits. Their methods were soon introduced into animal and plant breeding (e.g. Lush (1937)). But it took until the late 20th century (e.g. Bulmer (1980)) that the principles of quantitative genetics began playing an important role in evolutionary biology. With the rapid advances in molecular biology it became possible to actually identify loci underlying quantitative variation, an empirical return to the theoretical roots of quantitative genetics.

The early work in quantitative genetic theory and the need for quantitative methods to model and describe the distributions of continuously distributed characters led the development of modern statistical methodology. Francis Galton introduced the idea of regression and correlation by his *regression toward the mean* of parent and offspring measurements. Fisher (1918) introduced the concept of variance-component partitioning in order to separate the total phenotypic variance into additive, dominance, epistatic and environmental parts. Wright (1921a) developed the method of path analysis to analyze the inheritance of body characteristics in animals.

1.2 Model Selection

1.2.1 Concept

A conceptual model is a *representation* of some phenomenon, data or theory by logical and mathematical objects. In the case of statistical models, nondeterministic phenomena are considered. Suppose we have collected data, Y, and we have a number of competing models of different kind and complexity available to describe the data. We call the collection of candidate models $\mathcal{M}$. The model selection problem is now to choose – based on the data Y – a "good" model from the set of possible models $\mathcal{M}$.

Generally, a complex model will fit the data better than a simple model, but might do worse in extrapolating to related sets of data, i.e. another random sample Y^+ independent of Y, which seems to be a natural requirement.

These issues were already addressed by William of Ockham (1285-1347) by his *principle of parsimony* or Ockham's razor, which says: "entities should not be multiplied beyond necessity" or in other words: "it is in vain to do with more what can be done with fewer". Hence, simple models should be favored over complex ones that fit data about equally well. The critical point of his principle is however kept quite vague, namely the term "necessity" or the definition of what can be done with fewer and with more.

In statistical terms, the tradeoff between goodness-of-fit and simplicity can be interpreted as a compromise between bias and variance. The bias of the model will be larger for a simpler model while increasing the complexity of the model increases its variance.

Model selection employs some measure of optimality to choose between models of different classes and complexities.

1.2.2 Model selection criteria

There are two major classes of model selection criteria available. The first group aims at selecting the model with the largest posterior probability while the second approach tries to minimize the expected predictive error of the model. The most prominent and widely used members of the two classes are the Bayes Information Criterion (BIC, Schwarz (1978)) and the Akaike Information Criterion (AIC, Akaike (1973)), respectively. Both are applied in the following way: For a random sample Y of sample size n, we choose model M with p-dimensional parameter vector θ that attains the smallest value of the respective criteria. Both criteria consist of two terms, minus two times the log-likelihood of the data under the model plus a penalty term:

$$BIC(M) = -2 \log L(Y|M, \theta) + p \log n \qquad (1.1)$$

$$AIC(M) = -2 \log L(Y|M, \theta) + 2p \qquad (1.2)$$

The following sections give a detailed motivation of the two model selection criteria.

Bayesian information criterion

Suppose we have a set of candidate models $M_i, i = 1, \ldots, m$ and corresponding model parameters θ_i. The Bayesian approach to model selection aims at choosing the model with maximum posterior probability. Suppose $\pi(M_i)$ is the prior probability for model M_i and $f(\theta_i|M_i)$ is the prior distribution for the parameters of model M_i. Then the posterior probability of a given model is proportional to

$$P(M_i|Y) \propto \pi(M_i)P(Y|M_i) \qquad (1.3)$$

with

$$P(Y|M_i) = \int L(Y|\theta_i, M_i)f(\theta_i|M_i)d\theta_i \qquad (1.4)$$

Generally, the prior over models is assumed uniform, so that $\pi(M_i)$ is constant. This assumption will be relaxed in Section 1.3.2. The asymptotically relevant (for large n) terms in $P(Y|M_i)$ can be isolated using a Laplace approximation, leading to

$$\log P(Y|M_i) = \log L(Y|\hat{\theta}_i, M_i) - \frac{p_i}{2}\log 2\pi + \frac{1}{2}\log |H| + \mathcal{O}(n^{-1}) \qquad (1.5)$$

Here $\hat{\theta}_i$ is a maximum likelihood estimate, n is the sample size and p_i is the number of free parameters in model M_i. $|H|$ is the $p_i \times p_i$ Hessian matrix of $\log P(\theta_i|Y, M_i)$: $H = \frac{\partial^2}{\partial\theta\partial\theta^T}\log L(Y|\theta_i, M_i)|_{\hat{\theta}_i}$. For large sample sizes, the terms independent of n in Equation 1.5 can be dropped and $\log|H|$ can be approximated by $\frac{p_i}{2}\log n$. This leads us to:

$$\log P(M_i|Y) \approx BIC(M_i) = \log L(Y|\hat{\theta}_i, M_i) - \frac{p_i}{2}\log(n) \qquad (1.6)$$

Therefore, choosing the model with minimum BIC is (approximately) equivalent to choosing the model with largest posterior probability.

Akaike's information criterion

Suppose that the observed data Y are generated by an unknown true model with density function $f(Y)$. We try to find the closest candidate model $M_i, i = 1, \ldots, m$ with corresponding model parameters θ_i and probability density function $g(Y|\theta_i, M_i)$, by comparing $f(Y)$ and $g(Y|\theta_i, M_i)$. In the case of AIC, the distance is measured by the Kullback-Leibler distance:

$$D(f, g(\theta_i)) = \int f(y) \log \frac{f(y)}{g(y|\theta_i, M_i)} dy , \qquad (1.7)$$

which can be written as

$$D(f, g(\theta_i)) = \int f(y) \log f(y) dy - \int f(y) \log g(y|\theta_i, M_i) dy . \qquad (1.8)$$

Each of the integrals in Equation 1.8 is a statistical expectation with respect to the true distribution f. When comparing the Kullback-Leibler distance of two candidate models, the expectation $E_f \log f(Y)$ cancels out. Therefore, we write

$$D(f, g(\theta_i)) = C - E_f \log g(Y|\theta_i, M_i) . \qquad (1.9)$$

Equation 1.9 cannot be evaluated straight forward. $D(f, g(\theta_i))$ actually describes the distance between the true model and M_i with a specific parameter value θ_i. Therefore, θ_i is substituted by the value for which $D(f, g(\theta_i))$ obtains a minimum, which can be shown to be the MLE $\hat{\theta}_i$.

We can now compare the fit of two candidate models, M_1 and M_2, by taking the difference:

$$E_f \log g(Y|\hat{\theta}_2, M_2) - E_f \log g(Y|\hat{\theta}_1, M_1) .$$

Another problem becomes obvious when we consider a model M_1 that is nested within a more complex model M_2: $E_f \log g(Y|\hat{\theta}_2, M_2)$ will never be smaller than $E_f \log g(Y|\hat{\theta}_1, M_1)$. This happens because the same data is used, as so-called training data set, to estimate $\hat{\theta}_i$ and, as so-called validation data set, to assess the resulting fit.

AIC gives an estimate for the optimism of the model fit that arises when training and validation data set are identical by adding a penalty term (see equation 1.2) to the log-likelihood of each model. For a derivation of the penalty see e.g. Davison (2003), p. 150-152.

1.2.3 Model selection vs. Hypothesis Testing

Some important differences between model selection and hypothesis testing.

- Model selection usually involves many fits to the same set of data

- Hypothesis testing usually requires that the null hypothesis is chosen to be the simpler of the two models, i.e. that the models are nested. This is not necessarily true when comparing two candidate models.

- In model selection, we are not always assuming one model to be the true model. Ping (1997) states that "the existence of a true model is doubtful in many statistical problems. Even if a true model exists, there is still ample reason to choose simplicity over correctness knowing perfectly well that the selected model might be untrue. The practical advantage of a parsimonious model often overshadows concerns over the correctness of the model. After all, the goal of statistical analysis is to extract information rather than to identify the true model. The parsimony principle should be applied not only to candidate fit models, but the true model as well."

1.3 Mapping of Quantitative Trait Loci

As mentioned in Section 1.1.1, a quantitative trait is typically influenced by a number of interacting genes. In order to locate QTL, geneticists use molecular markers. These are pieces of DNA whose characteristics (i.e. genotypes) can be determined experimentally and that exhibit variation between individuals. Their . In organisms where chromosomes occur in pairs (i.e. diploid organisms), the genotype at a particular locus is specified by two pieces of DNA that are potentially different. If a QTL is located close to a given marker, we expect to see an association between the genotype of the marker and the value of the trait.

Locating QTL in natural, outbred, populations is relatively difficult. This is due to the fact that as a result of crossover events, which occur every time gametes are produced, the association between a QTL and a neighboring marker may be very weak. Therefore, to control the number of crossovers, scientists usually use data from families with more than one offspring or extended pedigrees (see e.g. Lynch and Walsh (1998) or Thompson (2000)). Locating QTL is easier and usually more precise when the data come from experimental populations. Such populations consist of inbred lines of individuals, who are homozygous at every locus on the genome (i.e. have identical pairs of chromosomes). By crossing individuals from such inbred lines, scientists can control the number of meioses (the process in which gametes are produced) and produce large experimental populations for which the correlation structure between the genotypes at different markers is easy to predict. Inbred lines have been created in many species of plant, as well as animal species (e.g mice). Results from research on experimental populations can often be used to predict biological phenomena in an outbred population, due to the similarity of genomes in the two populations (see e.g.

Phillips (2002)).

The work presented in this thesis deals exclusively with data from experimental crosses of well-defined strains of an organism. The process of inbreeding has fixed a large number of relevant traits in these strains. Therefore, if two strains, raised under similar environmental conditions, show consistent differences, we can assume a genetic basis of these differences.

In order to identify the genetic loci responsible for these differences, a series of experimental crosses between these two strains can be carried out. Two of the most common approaches are backcross and intercross designs (see Figure 1.1).

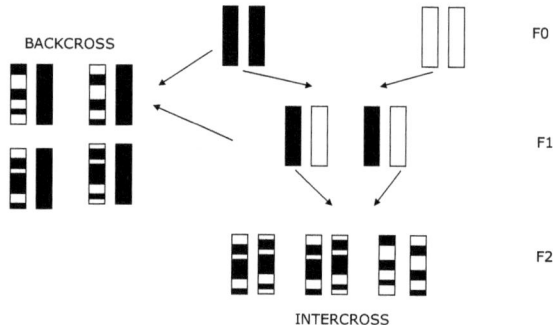

Figure 1.1: Types of experimental crosses. One pair of columns corresponds to the genome of one individual. The parents (F0–generation) have both identical pairs of chromosomes, maternal and paternal DNA is indicated by black and white bars, respectively. The children (F1) inherit one chromosome from their father and one from their mother. Backcross populations are achieved by crossing children with either father or mother. In intercross populations, grandchildren (F2) are produced.

1.3.1 QTL Mapping Techniques for Experimental Crosses

Approaches to QTL Mapping (see Figure 1.2) can be divided

- into univariate and multiple methods

- into marker based techniques and methods that estimate the QTL location

- by how they deal with epistasis, i.e. interacting, QTL

- into classical statistical or Bayesian methods

The extension of univariate to multiple methods is preferable for many reasons: increased power of detection, reduced bias of estimates of effect size and location, improved separability of correlated effects. This applies especially to QTL mapping, where marker data is typically non-orthogonal. There are, however, computational challenges due to the large number of possible models. In the case of marker based techniques, we typically deal with 50 to 500 possible predictors. The number can increase dramatically for methods that try to estimate the QTL location more precisely.

Interactions between QTL effects (i.e. epistasis) are a common phenomenon, which is supposed to play an important role in the genetic determination of complex traits (see e.g. Doerge (2002), Carlborg and Haley (2004) and references given there) as well as in the evolution process (see e.g. Wolf et al. (2000)). Neglecting these effects may lead to oversimplified models describing the inheritance of complex traits and to severely biased estimates of effects. Methodical, there are approaches that proceed hierarchically, i.e. they consider interactions only between already identified main effects. This reduces the computational complexity compared to searching for all possible interaction effects but causes an underestimation of the frequency and importance of epistasis (Wolf et al. (2000)) and fails to identify interesting regions of the genome.

A simple marker based univariate approach is to calculate a series of one-factorial ANOVAs comparing the mean phenotype of individuals with genotype "black"-"black" and "white"-"black" (and "white"-"white" in case of intercross designs) at each marker position ("black" and "white" refer to the colors used in Figure 1.1). Significant differences at particular marker positions indicate a QTL in the proximity of the marker. Classical interval mapping (Lander and Botstein (1989)) tries to give more precise estimates of the QTL location. For a dense grid of possible QTL positions, LOD (logarithm of odds) scores are derived that compare the evaluation of the likelihood function under the null hypothesis (no QTL) with the alternative hypothesis (QTL at the testing position).

Methods that try to extend interval mapping to a (pseudo-)multiple approach while keeping the computational intensity low include multiple interval mapping (MIM, Kao et al. (1999)), composite interval mapping (CIM, Zeng (1993, 1994)) and multiple interval mapping (MQM, Jansen (1993) and Jansen and Stam (1994)). MIM first locates all single QTL, then builds a statistical model with these QTL and their interactions and, finally, searches in one dimension for significant interactions. CIM and MQM perform interval mapping with a subset of marker loci as covariates in order to reduce the residual variation. The choice of suitable markers to serve as covariates could not be solved satisfactorily (see Broman (2001)).

A semi-Bayesian approach was developed by Sen and Churchill (2001). They first identify interesting regions on the genome using one- and two-dimensional LPD (log of posterior distribution) curves, based on multiple imputations of a pseudomarker grid. In a subsequent step, QTL from these interesting regions are chosen by applying a multiple regression model with standard model selection criteria.

Strict Bayesian approaches to QTL mapping were introduced by Yi and Xu (2002), Yi et al. (2003) and Yi et al. (2005). The methods are based on Markov Chain Monte Carlo (MCMC) algorithms that sample sequentially from QTL location, QTL genotype and genetic model. The number of QTL effects is allowed to change by Reversible-Jump MCMC. Generally, these methods are computationally involved and challenging to apply (see van de Ven (2004)).

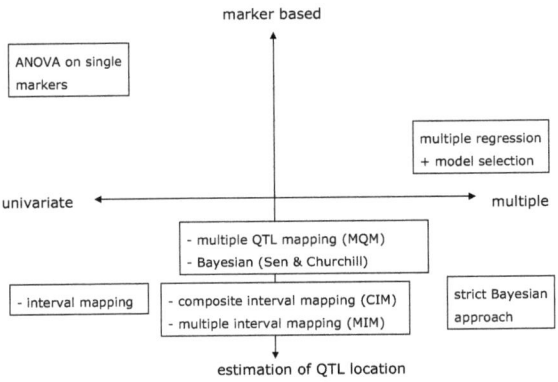

Figure 1.2: Overview of available techniques for QTL mapping

1.3.2 QTL Mapping by Model Selection

A multiple, marker based approach that allows for interaction effects independently of main effects involves fitting a multiple regression model relating the trait values to the genotypes of the markers. The most difficult part of this process is to estimate the number of QTL.

Model selection criteria can be employed to solve this problem. Here, we are less interested in the predictive power of the model than in finding the true minimal model. The class of the model is fixed as we are only considering linear models. Estimating the number of QTL corresponds to choosing the complexity of the model.

In the context of QTL mapping, the application of model selection criteria has been discussed e.g. by Broman (1997), Piepho and Gauch (2001), Ball (2001), Broman and Speed (2002), Bogdan et al. (2004) and Siegmund (2004). In particular Broman (1997) and Broman and Speed (2002) observed that the usually conservative BIC has a strong tendency to overestimate the number of QTL. This is maybe not unexpected since the BIC proposed by Schwarz (1978) is based on an asymptotic approximation using the Bayes rule to derive posterior probabilities for all the competing submodels of a regression model. The BIC cannot be expected to provide a good approximation in cases where the number of potential regressors is large compared to n.

To understand this phenomenon, notice that the arguments regarding the asymptotics which lead to the BIC imply that the prior is negligible and as a consequence all models are taken to be equally probable by the BIC. However, if the number of regressors is very large, then there are many more high dimensional models than low dimensional ones (when there are p^* potential regressors there are actually $\binom{p^*}{k}$ submodels of dimension k). As a consequence, it is likely that some of these higher dimensional models lead to a low value of the BIC just by chance.

The Modified BIC

Bogdan et al. (2004) propose a modified version of the BIC, the mBIC, which exploits the Bayesian context of the BIC and supplements this criterion with additional terms taking into account a realistic prior distribution on the number of QTL.

Assume n_m markers, i.e. potential regressors, are available. Considering all $n_e = n_m(n_m - 1)/2$ two-way interaction terms, we can construct $2^{n_m+n_e}$ different models. For assigning probabilities to these models, Bogdan et al. (2004) follow the standard solution proposed in George and McCulloch (1993): the ith main effect and jth interaction effect appear in the model with probabilities α and ν, respectively. For a particular model M involving p main effects and q interaction effects we obtain

$$\pi(M) = \alpha^p \nu^q (1-\alpha)^{n_m-p}(1-\nu)^{n_e-q} \ . \tag{1.10}$$

This choice implies binomial prior distributions on the number of main and interaction effects with parameters n_m and α, and n_e and ν, respectively. Substituting α by $1/l$ and ν by $1/u$ leads to

$$\log \pi(M) = C - p\log(l-1) - q\log(u-1) \tag{1.11}$$

In the context of multiple linear regression, minimizing the BIC (1.1) is equivalent to minimizing $n \log \text{RSS} + (p+q) \log n$. Hence the modified BIC, mBIC, chooses the model that minimizes the following quantity:

$$mBIC = n \log \text{RSS} + (p+q) \log n + 2p \log(l-1) + 2q \log(u-1) \ . \tag{1.12}$$

Prior information on the number of expected main (EN_m) and interaction effects (EN_e) can be used to assign the parameters $l := n_m/\text{EN}_m$ and $u := n_e/\text{EN}_e$. In the absence of prior information, Bogdan et al. (2004) propose the use of $\text{EN}_m = \text{EN}_e = 2.2$. This choice guarantees that the probability of a type I error under H_0 using the appropriate procedure (i.e. detecting at least one QTL when there are none) is smaller than 0.07 for sample sizes $n \geq 200$ and a moderate number of markers ($M > 30$).

Bogdan et al. (2004) and Baierl et al. (2006) present the results of an extensive range of simulations, which confirm the good properties of the standard and extended version of mBIC when applied to QTL mapping.

Chapter 2

Locating Multiple Interacting Quantitative Trait Loci in Intercross Designs

Based on:

Baierl, A., Bogdan, M., Frommlet, F. and Futschik, A. (2006). On Locating Multiple Interacting Quantitative Trait Loci in Intercross Designs. Genetics: 173,1693–1703.

2.1 Summary

A modified version (mBIC) of the Bayesian Information Criterion (BIC) has been previously proposed for backcross designs to locate multiple interacting quantitative trait loci. In this chapter, we extend the method to intercross designs. We also propose two modifications of the mBIC. First we investigate a two-stage procedure in the spirit of empirical Bayes methods involving an adaptive, i.e. data based choice of the penalty. The purpose of the second modification is to increase the power of detecting epistasis effects at loci where main effects have already been detected. We investigate the proposed methods by computer simulations under a wide range of realistic genetic models, with non-equidistant marker spacings and missing data. In case of large inter-marker distances we use imputations according to Haley and Knott regression to reduce the distance between searched positions to not more than 10 cM. Haley and Knott regression is also used to handle missing data. The simulation study as well as real data analysis demonstrate good properties of the proposed method of QTL detection.

2.2 Introduction

Consider a situation where we have a fairly densely spaced molecular marker map and our goal is to locate multiple interacting quantitative trait loci (QTL) influencing the trait of interest. We assume that marker genotype and quantitative trait value data are obtained by carrying out an intercross experiment using two inbred lines.

Due to the increased number of genotypes for the intercross design, the corresponding number of potential regressor variables describing additive and epistatic QTL effects is much larger than for the backcross design. We thus adapt the approach of Bogdan et al. (2004), and construct a modified version mBIC of the BIC for the intercross design. Additionally, we propose two new modifications of the BIC. The first of them is in the spirit of empirical Bayes approaches and is based on a two-step procedure. In the first step, the proposed mBIC is used for an initial estimation of the number of QTL and interactions. In the second step, QTL are located using the mBIC with the penalty modified according to the estimates obtained in step one. The second modification relies on extending the search procedure and is aimed at increasing the power of detection of interaction effects. We propose to consider an additional search for interactions which are related to at least one of the additive effects found in the original scan based on the mBIC. Restricting our attention to a limited set of interactions reduces the multiplicity of the testing problem and allows to use a smaller penalty for including interactions.

We perform an extensive simulation study verifying the performance of our method. In order to account for the more complicated model structure in the intercross design, the range of models considered in the simulations is substantially larger than in Bogdan et al. (2004). We also include models with non-equidistant and missing marker data. In situatons when the distance between markers is large, we use imputations according to Haley and Knott regression to keep the distance between searched positions smaller than or equal to 10 cM. We also investigate the use of Haley and Knott regression to handle missing data. Additionally, we apply our procedure to real data sets and compare the results to standard QTL mapping techniques. Our simulations as well as the analysis of real data suggest good properties of the proposed method and demonstrate that the proposed modifications of the mBIC may help to increase the power of QTL detection while keeping the proportion of false discoveries at a relatively low level.

2.3 Methods

2.3.1 Statistical Model

To model the dependence between QTL genotypes and trait values, we use a multiple regression model with regressors coded as described in Kao and Zeng (2002). This method of coding effects is known as Cockerham's model and involves an additive and a dominance effect for each QTL locus as well as effects modeling epistasis between two loci. With r QTL this leads to the following linear model:

$$y = \mu + \sum_{i=1}^{r} \alpha_i x_i + \sum_{i=1}^{r} \delta_i z_i + \qquad (2.1)$$
$$+ \sum_{1 \leq i < j \leq r} \gamma_{i,j}^{(xx)} w_{i,j}^{(xx)} + \sum_{i \neq j} \gamma_{i,j}^{(xz)} w_{i,j}^{(xz)} + \sum_{1 \leq i < j \leq r} \gamma_{i,j}^{(zz)} w_{i,j}^{(zz)} + \epsilon ,$$

where y is the trait value, and $\epsilon \sim N(0, \sigma)$ summarizes environmental effects. The variables are coded as specified below.

Additive Effects: $\quad x_i = x(g_i) = \begin{cases} 1 & \text{if } i^{th} \text{ QTL has genotype } g_i = A_i A_i, \\ 0 & \text{if } i^{th} \text{ QTL has genotype } g_i = a_i A_i, \\ -1 & \text{if } i^{th} \text{ QTL has genotype } g_i = a_i a_i. \end{cases}$

Dominance Effects: $\quad z_i = z(g_i) = \begin{cases} 1/2 & \text{if } i^{th} \text{ QTL has genotype } g_i = A_i a_i, \\ -1/2 & \text{else} . \end{cases}$

Epistatic Effects:
$$w_{i,j}^{(xx)} = w^{(xx)}(g_i, g_j) = x_i \cdot x_j ,$$
$$w_{i,j}^{(xz)} = w^{(xz)}(g_i, g_j) = x_i \cdot z_j ,$$
$$w_{i,j}^{(zz)} = w^{(zz)}(g_i, g_j) = z_i \cdot z_j .$$

The advantage of the Cockerham parametrization is that under linkage equilibrium, the effects are orthogonal and the coefficients α_i, δ_i and $\gamma_{i,j}$ have a natural genetic interpretation (see Kao and Zeng (2002)). The formulation of the model allows some of the coefficients to be zero to accommodate cases when there are either QTL that are not involved in epistatic effects, or QTL that do not have their own main effects yet influence the quantitative trait by interacting with other genes, i.e. epistatic effects.

If the experiment is based on a relatively dense set of markers, the first step in QTL localization could rely on identifying markers which are closest to a QTL. Thus our task reduces to choosing the best model of the form

$$y = \mu + \sum_{i \in I_1} \alpha_i x_i + \sum_{i \in I_2} \delta_i z_i + \\ + \sum_{(i,j) \in U_1} \gamma_{i,j}^{(xx)} w_{i,j}^{(xx)} + \sum_{(i,j) \in U_2} \gamma_{i,j}^{(xz)} w_{i,j}^{(xz)} + \sum_{(i,j) \in U_3} \gamma_{i,j}^{(zz)} w_{i,j}^{(zz)} + \epsilon ,$$
(2.2)

where I_1 and I_2 are certain subsets of the set $\mathcal{N} = \{1, \ldots, m\}$, m is the number of available markers, and U_1, U_2 and U_3 are certain subsets of $\mathcal{N} \times \mathcal{N}$. Analogous to the formulas given above, the values of the regressor variables x_i, z_i, $w_{i,j}^{(xx)}$, $w_{i,j}^{(xz)}$, $w_{i,j}^{(zz)}$ are defined according to the genotypes of the ith and j^{th} marker. Similarly to Bogdan et al. (2004), we allow interaction terms to appear in our model even when the related main effects are not included.

2.4 A Modified BIC for Intercross Designs

In this paper, we construct a version of mBIC suitable for intercross design. Note that in this design $m_v = 2m$ (m possible additive and m possible dominance terms), and $m_e = 2m(m-1)$. Choosing again $\mathrm{E}N_v = \mathrm{E}N_e = 2.2$, the resulting modified version of the BIC recommends the model for which

$$mBIC = n \log \text{RSS} + (p+q) \log n + 2p \log(m/1.1 - 1) + 2q \log(m(m-1)/1.1 - 1),$$
(2.3)

obtains a minimum. Here p is equal to the sum of the number of additive and dominance effects present in the model and q is the number of epistatic terms.

Observe that the proposed penalty for including individual terms is larger in the intercross design than in the backcross design. This is a result of a larger number of possible terms in the regression model, which forces us to increase the threshold for adding an additional term in order to keep control of the overall type I error. An upper bound for the type I error of the search procedure is derived using the Bonferroni inequality (see Appendix for details). Simulations show that the upper bound is close to the observed type I error for markers that are not closer than 5 cM.

Figure 2.1 compares the upper bound on the type I error of the mBIC when the penalty is adjusted for intercross designs (see Formula (2.3)) with the related type I error when the penalty designed for backcross ($2p \log(m/2.2 - 1) + 2q \log(m(m-1)/2.2 - 1)$) is used. The results are for $m = 132$ markers. The graph clearly shows that for common sample sizes adjusting the penalty is necessary to control the type I error at a 5% level.

Apart from using mBIC in its standard form (2.3), we developed adaptive strategies to modify the size of the penalty based on the data. In general, available prior information on the number of main and epistatic effects may be used to adjust the criterion in the following way:

$$mBIC_1 = n \log \text{RSS} + (p+q) \log n + 2p \log(2m/\text{E}N_v - 1) + 2q \log(2m(m-1)/\text{E}N_e - 1),$$
(2.4)

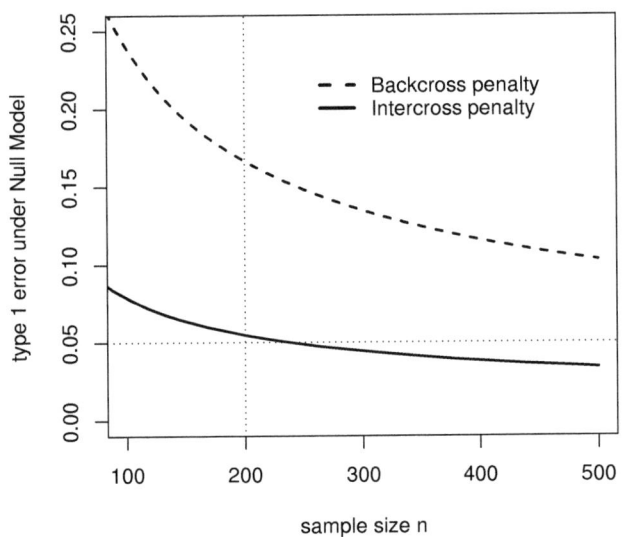

Figure 2.1: Comparison of the Bonferroni type I error bounds under the null model (no effects) for the intercross design when the same penalty as in backcross is used and when the penalty is adjusted accordingly

where EN_e and EN_v denote the expected values of N_e and N_v under the prior distribution. If we have no knowledge on the number of QTL, an obvious option is to use the data to obtain an initial estimation of N_e and N_v. Such estimates for N_e and N_v could in principle be obtained using standard methods for QTL localization, e.g. interval mapping. However, due to the known problems related to interval mapping (many local maxima between markers, difficulties with separating linked QTL and "ghost" effects) we recommend the application of the standard version of mBIC (2.3) for an initial search. We denote the number of additive and epistatic effects found in this initial search by $\hat{N}_v$ and $\hat{N}_m$. In the second step, the final localization of QTL is based on version (2.4) of the criterion, with EN_v replaced by $\max(2.2, \hat{N}_v)$ and EN_e replaced by $\max(2.2, \hat{N}_e)$.

In case of a large number of underlying QTL, the reduced penalty in the second search step increases the power of QTL detection. If in the first search step two or fewer main and epistatic effects are found, the penalty is not decreased. Thus in particular under the null model of no effects, the type I error will still be close to 5%.

We also consider a second extension to the search strategy in order to increase the power of detecting epistatic effects. The described application of the mBIC takes into account epistatic effects regardless of whether the corresponding main effects were included in the model or not. Therefore, epistasis can be detected in cases where main effects are weak or not present at all. Wolf et al. (2000) list the common practice of fitting epistatic terms after main effects have been included in the model as a main reason why in many QTL studies, epistasis has not been detected. However, the price for the possibility of detecting epistasis even if main effects are not detectable is a relatively large penalty for interaction terms. In particular for small sample sizes, this results in low detection rates (see Figure 2.2). This observation confirms the statement of Carlborg and Haley (2004) that epistatic studies "are most powerful if they use good quality data for 500 or more F_2 individuals".

For the above reasons, we deploy a third search step that increases the power of detection for epistatic terms by considering a restricted set of potential terms based on prior analysis. Specifically, we restrict our attention to those epistatic effects related to at least one of the main effects detected by an initial search based on (2.4). Thus the set of epistatic effects to be searched through in this third step consists of not more than $4p(m-1)$ elements, where p is the number of main

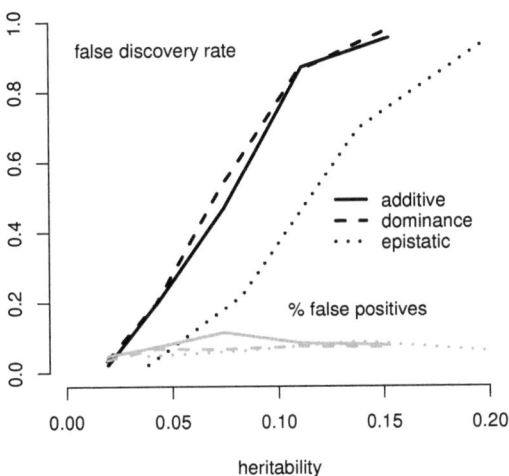

Figure 2.2: The dark curves show the percentage of correctly identified additive, dominance and epistatic effects depending on the heritability. The grey shaded curves display the expected number of incorrectly selected (linked and unlinked) markers ($n = 200$).

effects detected in the 2-step procedure. This allows us to decrease the penalty for interactions accordingly. The mBIC version used in this last step chooses the model which minimizes the quantity

$$mBIC_2 = n \log \text{RSS} + (p + q + q_a) \log n + 2p \log(2m/\text{EN}_v - 1) + \\ + 2q \log(2m(m-1)/\text{EN}_e - 1) + 2q_a \log(4p(m-1)/\text{EN}_e - 1) , \quad (2.5)$$

where q is the number of epistatic effects found in the 2-step procedure and q_a is the number of extra epistatic terms considered in the additional search for epistasis.

The penalty for the extra interaction terms in (2.5) is now of the same order as the penalty for additive terms and thus the power for detecting such epistatic effects should be comparable to the power of detecting main effects with the same heritability.

The identification of the model minimizing (2.3), (2.4) or (2.5) within the huge class of potential models is by no means trivial. Our approach is to use a forward selection procedure with the following stopping rule: if a local minimum of the modified BIC is reached, we still proceed with forward selection, trying to include (one-by-one) 5 additional terms. If, at some point, this leads to a new minimum, we temporarily accept this "best" model and continue again with forward selection. Otherwise none of the additional five effects are added. This approach helps to avoid premature stopping of the search algorithm at a local minimum. This can be the case when including two additional regressors improves the model even if each single one of them does not. The maximum number of additional regressors is set to five because it is very unlikely that five additional regressors improve the model while each of them alone does not or only marginally.

Finally, backward elimination is tried, i.e. it is checked whether mBIC can still be improved by deleting some of the previously added variables.

2.5 Simulations

Simulations are carried out to investigate the performance of our proposed method of QTL detection in the intercross design under a variety of parameter settings. All simulations were carried out in Matlab, the complete program is included in chapter ??.

We consider several scenarios involving equidistant markers that are relatively easy to analyze, and three realistic scenarios designed according to an actual QTL experiment described in Huttunen et al. (2004).

In our equidistant scenarios, we simulate QTL and marker genotypes on 12 chromosomes each of length 100 cM. Markers are equally spaced at a distance of 10 cM with the first marker at position 0 and the eleventh marker at position 100 of each chromosome. This leads to a total number of available markers m of 132 and the standard version of mBIC (2.3) becomes

$$mBIC = n \log \text{RSS} + (p+q) \log n + 9.56p + 19.33q \,.$$

Genome length and marker density are kept constant in all simulations and are in accordance with previous simulation studies (Piepho and Gauch (2001) and Bogdan et al. (2004)) in order to increase comparability.

Further details for the equidistant (both simple and more complex) scenarios, and the realistic scenarios are provided below. We simulated the trait data under different models of the form (2.1). In all simulations the overall mean μ and the standard deviation of the error term σ were set to be equal to 0 and 1 respectively. For each scenario and parameter setting, the simulation results are based on 500 replications.

Among the simulation results we include are the average number of correctly identified effects, which we denote by c_{add}, c_{dom} and c_{epi} for additive, dominance and epistatic effects respectively. In the case of simple models with just one effect, these quantities are estimates of the power. A main effect is classified to be correctly identified, if the regression model chosen by mBIC includes the corresponding effect related to a marker within 15 cM of QTL. An epistatic effect is classified as correctly identified when the mBIC finds a corresponding effect with both markers falling within 15 cM of the corresponding QTL. If more than one effect is detected in such a window, only one of them is classified as true positive. All the other effects are considered to be false positives.

In our simulation study of more complex equidistant scenarios, we simulated many QTL with weak effects. In this situation, the confidence intervals for the estimates of QTL location are often much wider than 30 cM (see e.g. Bogdan and Doerge (2005)). Thus each of such weak effects will bring a certain proportion of "false" positives related to a weak precision of QTL localization, while still providing an approximation to the best regression model. As a result of this phenomenon, the total number of false positives typically increases with the size of the model used in the simulation. Therefore, additionally to the average number of false positives fp, we report the estimated proportion of false positives within the total number of identified effects, $pfp = fp/(c_{add} + c_{dom} + c_{epi} + fp)$.

Simple equidistant scenarios: We first consider the null model, i.e. the situation where there are no QTL at all. As shown in Figure 2.1, the probability that at least one effect is incorrectly selected should be below 0.05 when the sample size is at least 200. Our simulations lead to a percentage of 0.038 of such (familywise) type I errors when $n = 200$, thus confirming the theoretical results. The percentage of errors should decrease with increasing sample size, and indeed for a sample size of $n = 500$, the number goes down to 0.02.

Next we consider two experiments to investigate the detectability of QTL effects depending on their strength, effect type (additive, dominance or epistatic) and on the total number of QTL. In these experiments we use a sample size of $n = 200$.

For the first experiment, we generate the data according to three simple models of the form (2.1). In the first two models (scenarios 1 and 2), one QTL is located at the fifth marker on the first chromosome. In scenario 1 the QTL has only an additive effect with the effect size α ranging from 0.2 to 0.6. In scenario 2, the additive effect is constant ($\alpha = 0.7$) and a dominance effect δ with values in the interval between 0.4 and 1.2 is added. For scenario 3 only one epistatic effect ($\gamma_{1,2}^{(xx)}$) between markers number five of chromosome five and six respectively is considered. The effect size of ($\gamma_{1,2}^{(xx)}$) ranges between 0.4 and 1.6.

In the context of scenarios 1 and 3, we investigate the power of detection in dependence on the classical heritability

$$\frac{\sigma_*^2}{1+\sigma_*^2}, \qquad (2.6)$$

with 1 being the environmental variance, and σ_*^2 denoting the variance due to the single genetic effect present (i.e. $\sigma_*^2 = \sigma_{add}^2$ in case of an additive effect, and $\sigma_*^2 = \sigma_{epi}^2$ in case of epistasis between two loci).

In scenario 2, the power of detection of the dominance effect should also depend on whether the corresponding additive effect can be detected, since the error variance gets smaller, if the additive effect is included into the regression model. In our experiment the additive effect was almost always detected (power 99%) and we observed that a good indicator for the power of detection of the dominance term is its heritability in the model without the additive term

$$h_{dom}^2 = \frac{\sigma_{dom}^2}{1+\sigma_{dom}^2} = \frac{0.25\delta}{1+0.25\delta}. \qquad (2.7)$$

A comparison of detection rates of additive, dominance and epistatic effects in dependence on the heritability (as defined in (2.6) for additive and epistatic effects, respectively and in (2.7) for the dominance effect) is given in Figure 2.2. The relationship can be seen to be S-shaped and nearly identical for additive and dominance effects. Although dominance and additive effects are detected with the same power at a fixed heritability, the actual size of the dominance effects has to be larger (by $\sqrt{2}$) than the additive effects ($\sigma_{add}^2 = a^2/2$ and $\sigma_{dom}^2 = d^2/4$ for an additive effect of size a and dominance effect of size d. Hence if $\sigma^2_{add} = \sigma^2_{dom}$, d has to be $a\sqrt{2}$). For epistatic effects, the power of detection is lower. This can be explained by the increased penalty of the model selection criterion.

The grey shaded curves in Figure 2.2 display the average number of falsely detected effects, which can be used as an estimate of the expected number of false positives. This quantity is an upper bound to the probability of having at least one incorrect effect in the model. The displayed error rates are fairly constant over the range of heritabilities considered. They vary between 0.05 (the value achieved by the model with no effects) and 0.11.

The purpose of the second experiment is to investigate to what extent the power of detection of individual signals is affected by the amount of QTL influencing the trait. The number of QTL varies between one and 10, all QTL are on different chromosomes and therefore unlinked and have only additive effects with $\alpha_i = 0.5$.

Figure 2.3 shows that the probability of detection using the standard version of mBIC (2.3) decreases with the number of effects present. This can be explained by the fact that criterion (2.3) is based on the assumption that the expected number of effects is equal 2.2. If the correct (but in practice unknown) number of effects were used instead of 2.2, the percentage of correctly identified additive terms would increase from 0.543 to 0.761 for 10 underlying effects, from 0.672 to 0.781 for 7 and from 0.763 to 0.7995 for 4 underlying effects. We can obtain a comparable improvement by applying the two step procedure defined in Formula (2.4) that involves an estimation of the number of expected effects in the first search step. The dotted line in Figure 2.3 shows that the power of detection increases while the proportion of false positives remains stable.

Complex equidistant scenarios: Here, we consider nine more complex models that involve several effects of different size and type.

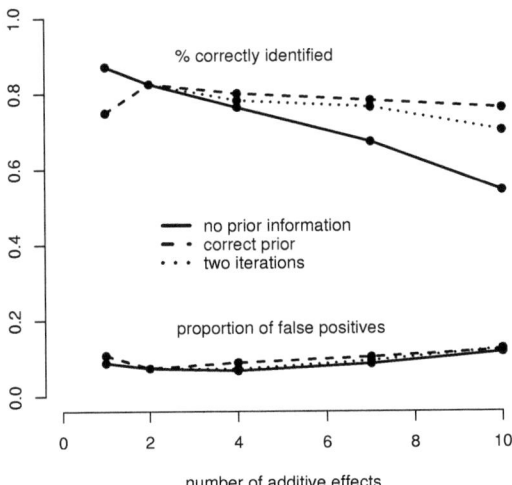

Figure 2.3: Percentage of correctly identified additive effects vs. number of additive effects. The QTL are unlinked, i.e. located on different chromosomes, and have effect sizes of 0.5. The solid line is based on simulations where no prior information is used to derive the penalty terms of the modified BIC. The dashed line represents simulations with the correct number of underlying effects (1,2,4,7,10) assumed known. The dotted line corresponds to the two step search procedure based on Formula (2.4).

For all models, the overall broad sense heritability $h_b^2 = \sigma_G^2/(\sigma_\varepsilon^2 + \sigma_G^2)$ is kept at 0.7; i.e 70 % of the phenotypic variance is explained by genotypic variation σ_G^2. Fixing the variance caused by environmental effects σ_ε^2 to 1 leads to a genotypic variation of 2.3̇, which is then distributed among additive effects (45%), dominance effects (25%) and epistatic effects (30%). The resulting narrow sense heritability has an expected value of 0.315. All simulations are done both with sample size 200 and 500.

We consider all combinations of situations involving two, four and eight additive and epistatic effects. Dominance effects are assigned to half of the loci where additive effects occur. The epistatic QTL are taken both from the additive effect positions and from other genome locations. If p additive effects are present, the relative size of effect i is chosen to yield $100\frac{i}{p(p+1)/2}\%$ of the additive heritability. For dominance and epistatic effects the relative strengths are chosen analogously. We consider the worst case situation where the QTL positions are always exactly in the middle of two markers. Table 2.1 contains a brief summary of the resulting nine scenarios. A detailed description of all effect positions and strengths can be found on our web page *http://homepage.univie.ac.at/andreas.baierl/pub.html* .

Table 2.1: Description of scenarios 1-9

scenario	n_{add}	n_{dom}	n_{epi}^{aa}	n_{epi}^{dd}	n_{epi}^{ad}
1	2	1	1	1	0
2	2	1	3	0	1
3	2	1	7	1	0
4	4	2	1	1	0
5	4	2	3	0	1
6	4	2	7	1	0
7	8	4	1	1	0
8	8	4	3	0	1
9	8	4	7	1	0

Columns contain number of additive (n_{add}), dominance (n_{dom}) and epistatic QTL for each scenario. Epistatic effects can be of additive-additive (n_{epi}^{aa}), dominance-dominance (n_{epi}^{dd}) or additive-dominance (n_{epi}^{ad}) type.

Results for simulations with sample sizes of 200 and 500 are described in the following. Table 2.2 summarizes the average number of correctly identified effects as well as the average number of false positives and the proportion of false positives for the standard version of the mBIC (2.3). Table 2.3 gives the corresponding statistics for modifications based on the two step procedure (see Formula (2.4)) as well as for the additional search for epistatic terms with reduced penalty based on Formula (2.5). Table 2.3 demonstrates that the two step procedure has the potential to increase the detection power while keeping the observed proportion of

Table 2.2: Simulation results for sample size 200 and 500 (initial penalty)

scenario	1	2	3	4	5	6	7	8	9
c_{add}	1.818	1.712	1.668	2.424	2.334	2.182	2.950	2.676	2.478
c_{add}^*	1.994	1.988	1.982	3.180	3.166	3.158	4.982	4.976	4.874
c_{dom}	0.966	0.968	0.978	1.296	1.204	1.064	1.348	1.164	0.972
c_{dom}^*	0.932	0.962	0.958	1.814	1.870	1.840	2.626	2.260	2.530
c_{epi}	1.040	0.860	0.470	1.020	0.830	0.390	0.900	0.650	0.250
c_{epi}^*	1.730	2.350	3.690	1.730	2.260	3.280	1.630	1.990	3.110
pfp	0.040	0.053	0.063	0.050	0.044	0.059	0.061	0.065	0.080
pfp*	0.027	0.024	0.024	0.020	0.023	0.026	0.023	0.019	0.029
fp	0.160	0.200	0.210	0.250	0.200	0.230	0.340	0.310	0.320
fp*	0.130	0.130	0.160	0.140	0.170	0.220	0.220	0.180	0.310

Columns contain the average number of correctly identified additive (c_{add}), dominance (c_{dom}) and epistatic (c_{epi}) effects. "pfp" denotes the proportion of false positives and "fp" the average number of falsely detected effects, respectively. Without the superscript "*", the results are for a sample size of 200, otherwise they are based on samples of size 500.

false positives at a level similar to the standard version of the mBIC. The increase in detection rates gained by this procedure is apparent for models with a larger number of underlying QTL (scenario 7-9). The performance of the additional search for epistasis based on (2.5) depends on the actual model. In some cases the corresponding increase in the number of false positives is larger than the increase in the average number of correctly identified effects. We observed this situation to occur under scenarios 1,4 and 7 (one relatively weak epistatic effect related to one of the main effects) and the sample size $n = 200$. Notice however that under all these scenarios the gain in the detection rate was decisively larger than the increase in false positives when the sample size was $n = 500$. The additional search for epistatic effects is especially successful for scenario 9 with a large number of underlying main and epistatic effects.

Figures 2.4 and 2.5 are based on the final search results described in Table 2.3. They indicate that the ability to detect an effect of a given size depends mainly on the individual effect heritability $h^2 = \sigma_{eff}^2/(\sigma_\epsilon^2 + \sigma_G^2)$.

Table 2.3: Simulation results for sample size 200 and 500 (adjusted penalty)

scenario	1	2	3	4	5	6	7	8	9
c_{add}	1.822	1.700	1.694	2.494	2.456	2.353	3.086	3.014	2.600
c_{add}^*	1.993	1.990	1.995	3.280	3.238	3.218	5.200	5.163	5.140
c_{dom}	0.978	0.987	0.966	1.302	1.240	1.180	1.504	1.280	1.032
c_{dom}^*	0.938	0.945	0.943	1.865	1.848	1.865	2.833	2.390	2.755
c_{epi}	1.064	0.940	0.482	1.016	0.806	0.380	0.962	0.630	0.252
c_{epi}^*	1.745	2.365	3.780	1.685	2.220	3.410	1.625	1.963	3.278
fp	0.156	0.200	0.228	0.260	0.262	0.280	0.370	0.336	0.346
fp*	0.125	0.150	0.230	0.205	0.205	0.260	0.233	0.218	0.270
pfp	0.039	0.052	0.068	0.051	0.055	0.067	0.062	0.064	0.082
pfp*	0.026	0.028	0.033	0.029	0.027	0.030	0.024	0.022	0.024
$\Delta c_{a.epi}$	0.018	0.130	0.158	0.046	0.112	0.100	0.058	0.128	0.100
$\Delta c_{a.epi}^*$	0.150	0.040	0.070	0.185	0.035	0.085	0.145	0.110	0.395
Δfp_a	0.07	0.11	0.104	0.082	0.09	0.08	0.066	0.086	0.084
Δfp_a^*	0.065	0.04	0.065	0.035	0.025	0.055	0.05	0.027	0.048
pfp_a	0.055	0.074	0.088	0.065	0.069	0.080	0.071	0.075	0.095
pfp_a^*	0.038	0.034	0.042	0.033	0.030	0.035	0.028	0.025	0.027

Simulation results for the two step procedure based on Formula (2.4) and the three step procedure based on Formula (2.5) are shown. Columns contain the average number of correctly identified additive (c_{add}), dominance (c_{dom}) and epistatic (c_{epi}) effects as well as the average number of falsely detected effects (fp) and the proportion of false positives (pfp) for the two step procedure. The column $\Delta c_{a.epi}$ and Δfp_a display average numbers of correctly identified and false positive epistatic effects that were detected additionally in the third search step based on Formula (2.5); "pfp_a" on the other hand denotes the total proportion of false positives based on the finally selected model at the end of the third search step. Without the superscript "*", the results are for a sample size of 200, otherwise they are based on samples of size 500.

For the sample size $n = 200$, the majority of large additive effects ($h^2 > 0.07$) is detected with a high power (larger then 0.8). While only some fraction of moderate effects ($h^2 \in (0.04, 0.07)$) is detected for $n = 200$, moderate additive and dominance effects are almost always detected when $n = 500$. Epistasis effects are somewhat harder to detect, the type of epistasis however does not influence the detectability. The observed proportion of false positives never exceeds 9% for $n = 200$ and 4% for $n = 500$.

Realistic scenarios: As an alternative model, we take the marker setup from a Drosophila experiment by Huttunen et al. (2004) and also include missing data.

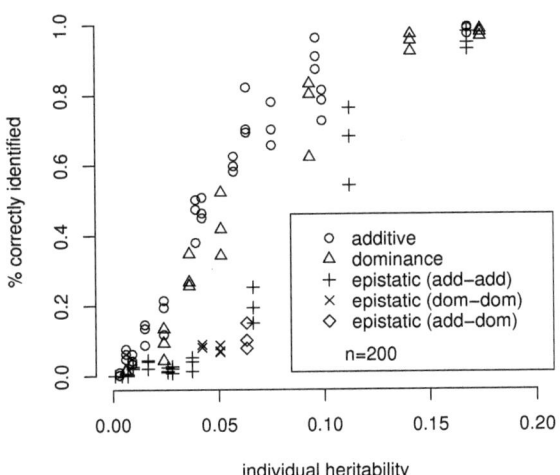

Figure 2.4: Percentage correctly identified additive, dominance and epistatic effects vs. individual effect heritabilities $h^2 = \sigma^2_{eff}/(\sigma^2_\epsilon + \sigma^2_G)$. Detection rates are taken from simulations of scenarios 1-9 (see Table 2.3) for $n = 200$.

To obtain a more densely spaced set of genome locations, genotype values were imputed at 35 positions chosen equidistantly between adjacent markers, keeping the maximum distance between the considered genome locations at not more than 10 cM. Haley-Knott regression (Haley and Knott (1992)) was used to impute values. See Figure 2.6 for the marker locations.

Our three scenarios permit for different expected proportions (0%, 5%, and 10% resp.) of marker locations per chromosome where the genotype information is missing. To permit for comparison, both heritabilities and QTL characteristics are chosen as in the above mentioned complex equidistant scenario 4 involving four additive, two dominance and two epistatic effects, and furthermore the QTL

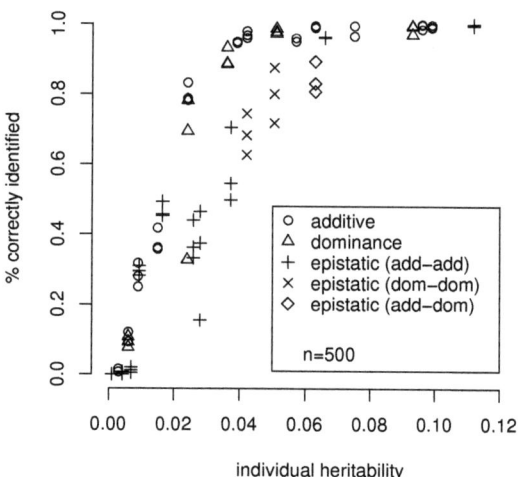

Figure 2.5: Percentage correctly identified additive, dominance and epistatic effects vs. individual effect heritabilities $h^2 = \sigma^2_{eff}/(\sigma^2_\epsilon + \sigma^2_G)$. Detection rates are taken from simulations of scenarios 1-9. (see Table 2.3) for $n = 500$.

effects have again been positioned in a distance of 5 cM to the closest marker. For this experiment we use the sample size $n = 200$.

According to Table 2.4, the obtained results are similar to those obtained for the complex equidistant scenario 4 which has the same number and relative strength of effects. This suggests that our approach does not rely on the somewhat unrealistic assumption of equidistant markers and no missing data.

Not surprisingly, the average number of correctly identified markers decreases slightly when the proportion of missing data increases. The proportion of false positives on the other hand somewhat increases. This results from a loss of power as well as a loss of precision in localizing QTL.

2.6 Illustrations

We apply our proposed method to data sets from QTL experiments on *Drosophila virilis* and mice, respectively. Huttunen et al. (2004) analyzed the variation in male courtship song characters in *Drosophila virilis*. We considered their intercross data set obtained from 520 males and the quantitative trait PN (number of pulses in a pulse train). Figure 2.6 shows the positions of the markers used in this experiment (solid lines). Depending on the chromosome, between two and five percent of the marker data were missing. We used the same imputation strategy as for our considered realistic scenarios, both for the missing data and the additional genome positions.

Huttunen et al. (2004) used single marker analysis as well as composite interval mapping. They found one QTL on chromosome 2, five QTL on chromosome 3 and another QTL on chromosome 4. As they note, four of the five positions found on chromosome 3 are close together and may well correspond to only one single underlying QTL.

Table 2.4: **Simulation results for different percentages of missing data**

missing	c_{add}	c_{dom}	c_{epi}	pfp
0%	2.486	1.302	1.290	0.043
5%	2.362	1.264	1.140	0.054
10%	2.262	1.160	0.980	0.070

Results of simulations for the realistic scenario with 0%, 5% and 10% of missing marker data. Columns contain the average number of correctly identified additive (c_{add}), dominance (c_{dom}) and epistatic effects (c_{epi}) and the proportion of false positives (pfp) derived with the two step search strategy based on Formula (2.4).

With our approach and the penalization based on 59 search positions, we found the same QTL positions on chromosome 2 (at 53.7 cM) and 4 (at 100.2 cM), but only two positions (at 25.4 and 108.25 cM) on chromosome 3. All QTL found

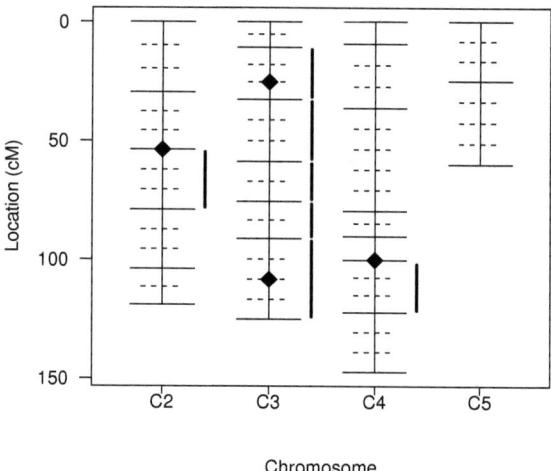

Figure 2.6: Genetic map for the *Drosophila v.* experiment by Huttunen et al. (2004). Solid horizontal lines indicate observed marker positions, dotted lines show imputed positions. QTL localized by our proposed method are symbolized by diamonds. Intervals with significant additive and/or dominance effects found by Huttunen et al. (2004) applying composite interval mapping are indicated by solid vertical lines.

were classified as additive. The QTL locations pointed out by our method as well as intervals suggested by Huttunen et al. (2004) are presented in Figure 2.6. In the results of the additional regression analysis we observed that none of the putative QTL suggested by Huttunen et al. (2004) on chromosome 3 that were not found by our method significantly improves our model (corresponding p-values for adding these QTL were equal to 0.85, 0.34, 0.06 and 0.32). Given these results and the above remark by Huttunen et al. (2004)), our method might have lead to a more precise localization of the respective QTL on chromosome 3.

Shimomura et al. (2001) investigated the circadian rhythm amplitude in mice on 192 F2 individuals. Genotypes were observed on 121 markers spread across the 19 autosomal chromosomes with 0-5% of the data missing for most markers. Again, the same imputation strategy as described in the section on realistic scenarios was used.

The analysis presented in Shimomura et al. (2001) consists of single and pairwise marker genome scans with permutation tests for assessing statistical significance. They identified one main effect on chromosome 4 at 42.5 cM and one epistatic term between the previous position and a marker on chromosome 1 at 81.6 cM.

Both QTL were detected by our method, one additive main effect on chromosome 4 and one dominant × additive epistatic effect between the QTL on chromosome 4 and 1. The epistatic term was found in an additional search step based on the mBIC described in Formula (2.5).

2.7 Discussion

In this paper we use a modification of the Bayesian Information Criterion (mBIC) to locate multiple interacting quantitative trait loci in intercross designs. The proposed procedure allows to detect multiple interacting QTL while controlling the probability of the type I error at a level close to 0.05 for sample sizes $n \geq 200$. The main advantages of this procedure include that it is straightforward to apply and computationally efficient which makes an extensive search for epistatic QTL practically feasible.

We presented results from simulations with single effects (additive, dominance and epistatic) of different magnitude and for complex scenarios in order to investigate detection thresholds. We applied our proposed procedure to realistic parameter settings including non-equidistant marker positions and different proportions of missing values. In order to demonstrate the applicability of our proposed method when applied to real data, we analyzed two sets of QTL experiments from the genetic literature, namely one dealing with Drosophila *virilis* and another one with mice. Both, our simulation results and our real data analysis confirm good properties of the proposed modifications to the BIC.

Compared to the original BIC (see Schwarz (1978)), the mBIC contains an extra penalty term which accounts for the large number of markers included in typical genome scans and the resulting multiple testing problem. While a modification of the BIC is already required when only main effects are considered (see e.g. Broman and Speed (2002) and theoretical calculations in Bogdan et al. (2004), the multiple testing problem becomes even more important when epistatic effects can enter the model regardless of the related main effects, as in our approach.

The simulations reported in this paper show that the mBIC appropriately separates additive, dominant and epistatic effects. In the case of closely linked markers however, our approach sometimes leads to a misclassification of the effect type, while still correctly identifying the presence of an effect. Hence, we suggest to use the mBIC rather to locate QTL than to identify the specific effect type. The procedure should also not be extended to estimate the magnitude of QTL effects or heritabilities. Estimating parameters after model selection leads to upward biased estimators of the effect sizes. This is true for any method leading to the choice of a single set of regressors, i.e. also in the case of the widespread methods based on multiple tests or interval mapping (see e.g. Bogdan and Doerge (2005)).

The prior for the number of main and epistatic effects in the standard version of the mBIC (with expected values $EN_v = 2.2$ and $EN_e = 2.2$) allows to control the probability of the type I error. This is suggested for an initial search in case of no prior knowledge on the number of effects. When reliable information on the number of effects is available, we strongly recommend using it for defining EN_v and EN_e. Our simulations also show that modifying the prior choices of EN_v and EN_e in Formula (2.4) using estimates of the QTL number from an initial search based on the standard version of the mBIC allows for some increase of power of QTL detection while preserving the observed proportion of false positives at a level similar to the standard version of the mBIC. The same holds for the additional search for epistatic terms with modified penalties according to Formula (2.5).

In the present paper we apply the mBIC to locate multiple interacting QTL by choosing the best of competing regression models. Our simulations as well as results reported in Broman (1997), Broman and Speed (2002) and Bogdan et al. (2004) show that the proposed forward selection strategy performs very well in this context. However, the mBIC has also great potential to be used in a stricter Bayesian context. The majority of currently used Bayesian Markov Chain Monte Carlo methods for QTL mapping requires multiple generation of all regression parameters and multiple visits of a given model in order to estimate its posterior probability by the frequency of such visits (see eg. Yi et al. (2005) and references given there). As a result, the proposed methods are computationally intensive and are very rarely verified by thorough simulation studies, which could provide insight into the influence of the prior distributions. The influence of the choice of priors on the outcome of Bayesian model selection methods is discussed e.g. by Clyde (1999). Note that the mBIC provides a method to estimate the posterior probability of a given model by visiting this model just once. This is because $exp(-mBIC/2)$ is an asymptotic approximation for $P(Y|M) * P(M)$, where $P(Y|M)$ stands for the likelihood of the data given model M (see Schwarz (1978)) and $P(M)$ is the prior probability of a given model. Thus the posterior probability of a given model M_i could be estimated by

$$P(M_i|Y) \approx \frac{\exp(-mBIC_i/2)}{\sum_{j=1}^{k}\exp(-mBIC_j/2)},$$

given that the k visited models contain all plausible models. To reach all sufficiently plausible models, a suitable search strategy needs to be designed. The construction of such an efficient search strategy is difficult due to the huge number of possible models (for 200 markers we potentially have 2^{638800} models). However, we believe that a numerically feasible procedure permitting to use mBIC in a Bayesian context might be found by exploiting the specific structure of QTL mapping problems, restricting the search space and applying a proper adaptation of an efficient MCMC sampler (see e.g. Broman and Speed (2002)) or a heuristic search strategy like genetic algorithms (see e.g. Goldberg (1989)), simulated annealing (Kirkpatrick et al. (1983)), tabu search (Glover (1989a), Glover (1989b)) or ant colony optimization (see e.g. Dorigo et al. (1999)). This would allow to estimate posterior probabilities of different plausible models as well as to use model averaging to estimate parameters like effect sizes and heritabilities.

2.8 Appendix

The difference between the mBIC of the null model ($mBIC_0$) and the mBIC of any one-dimensional model M_i ($mBIC_{M_i}$) is $\log n + 2\log(l-1)$ or $\log(u-1)$ depending whether the effect included in the one-dimensional model is a main or epistatic effect. The number of possible one-dimensional models M_i for intercross designs is $2m + 4m(m-1)/2$.

In order to derive a bound for the type I error under the null model, we note that two times the difference of the likelihoods of a one-dimensional model and the null model is approximately χ^2-distributed with 1 d.f.

Applying the Bonferroni inequality gives

$$P(mBIC_{M_i} > mBIC_0, \text{ for any } i) \leq 4mP(Z > \sqrt{\log n + 2\log(l-1)}) +$$
$$+ 4m(m-1)P(Z > \sqrt{\log n + 2\log(u-1)}) + \epsilon \quad (2.8)$$

for the probability of choosing any one-dimensional model, if the null model is true.

The curves shown in Figure 2.1 are derived by evaluating the right hand side of Equation 2.8 for values of n between 100 and 500 and $m = 132$. For the backcross penalty, the parameters l and u are set to $m/2.2$ and $m(m-1)/4.4$, respectively, whereas for the intercross penalty to $u = m/1.1$ and $l = m(m-1)/1.1$.

Chapter 3

Locating multiple interacting quantitative trait loci using robust model selection

Based on:

Baierl, A., Futschik, A., Bogdan, M. and Biecek, P. (2007).Locating multiple interacting quantitative trait loci using robust model selection. *Computational Statistics and Data Analysis* **51**, 6423-6434

3.1 Introduction

The mBIC is based on standard L_2 regression, assuming that the conditional distribution of the trait given the marker genotypes is normal. In practice this assumption is rarely satisfied. While the Central Limit Theorem shows that moderate deviations from normality have little influence on the mBIC, the properties of this criterion deteriorate drastically when the distribution of the trait has a heavy tail or the data include a certain proportion of outliers. Thus we consider an alternative approach based on robust regression techniques and construct robust versions of the mBIC. For this purpose, we use several well known contrast functions and investigate the resulting versions of the mBIC both analytically and using computer simulations. It turns out that the robust versions of the mBIC perform consistently well for all the distributions analyzed and much better than the standard version of the mBIC in situations when the distribution of the trait is heavy tailed. A possible exception to this rule is Huber's contrast function with a very small k, which we considered as a close approximation to L_1 regression. The corresponding procedure was outperformed by the other considered procedures under several models describing the error. While the basic idea of including a prior that penalizes high dimensional models more heavily should be of interest in a more general context, the modification of the BIC must be adapted to the structure of the model. Thus our investigations are carried out in an ANOVA setting with one way interactions, as encountered in the context of QTL mapping. Consequently, our regressors are taken to be the dummy variables which describe the genotypes of the markers, as well as products of pairs of these dummy variables.

3.2 The statistical model

We start by briefly reviewing the multiple regression model considered. It is an ANOVA model with one-way interactions and can be used for QTL mapping within the context of the backcross design. Let X_{ij} denote the genotype of the ith individual at the jth marker. We set $X_{ij} = -\frac{1}{2}$, if the ith individual is homozygous at the jth marker and $X_{ij} = \frac{1}{2}$, if it is heterozygous. We fit a multiple regression model of the form

$$Y_i = \mu + \sum_{j \in I} \beta_j X_{ij} + \sum_{(u,v) \in U} \gamma_{uv} X_{iu} X_{iv} + \epsilon_i \; , \tag{3.1}$$

where I is a subset of the set $N = \{1, \ldots, n_m\}$, n_m denotes the number of markers available, U is a subset of $N \times N$ and ϵ_i is a random error term. In order to identify markers which are close to a QTL, we need to select an appropriate submodel. For this purpose, we allow the inclusion of interaction terms in the model, even when the corresponding main effects are not included. This approach is justified by the recent discoveries of genes that do not have their own additive effects, but only influence a trait by interacting with other genes (see e.g. Fijneman et al. (1996) and Fijneman (1998)).

For normally distributed errors, ϵ_i, classical least squares regression is well justified, for instance, by the Gauss–Markov Theorem. As a result of the Central Limit Theorem, if the sample is large enough, then the standard test procedures for the significance of regression coefficients are resistant to moderate deviations from the assumption of normality. It can be seen that this property is shared by the BIC and its modification for mapping QTL, the mBIC. However, the estimates and tests derived under the assumption of normality cannot be expected to work well in the cases of skewed and heavy tailed distributions. In some situations, as in the case of the Cauchy distribution, the Central Limit Theorem does not even hold. It is also well known that standard L_2 regression is highly sensitive to outliers.

Methods of robust regression provide an alternative in such situations. They perform well under a wide range of error distributions without losing too much power when normality holds. One approach to obtaining robust regression estimates is to use M-estimates, i.e. to minimize another measure of distance instead of the residual sum of squares. Another approach would be to use MM-estimates, which have the additional property of also being robust with respect to leverage points (see Yohai (1987)). However, the model considered only contains dummy variables and, due to appropriate randomization, the design is close to being balanced. Therefore, leverage points should not be expected. We thus focus on M-estimates but mention here that our simulations gave nearly identical results for both M-estimates and the corresponding MM-estimates.

M-estimates of the regression parameters are based on the minimization of $\sum_{i=1}^{n} \rho(r_i)$, where the r_i are the residuals standardized using a robust scale estimator and $\rho(x)$ is a contrast function. We consider the following popular contrast functions:

$$\rho_{Huber}(x) := \begin{cases} k|x| - k^2/2 & \text{for } |x| > k \\ x^2/2 & \text{for } |x| \leq k \end{cases} \quad (3.2)$$

$$\rho_{Bisquare}(x) := \begin{cases} k^2/6 & \text{for } |x| > k \\ \frac{k^2}{6}[1 - (1 - (\frac{x}{k})^2)^3] & \text{for } |x| \leq k \end{cases} \quad (3.3)$$

$$\rho_{Hampel}(x) := \begin{cases} a(b-a+c)/2 & \text{for } |x| > c \\ a(b-a+c)/2 - \frac{a(|x|-c)^2}{2(c-b)} & \text{for } b < |x| \leq c \\ a|x| - a^2/2 & \text{for } a < |x| \leq b \\ x^2/2 & \text{for } |x| \leq a \end{cases} \qquad (3.4)$$

The calculation of regression coefficients based on these contrast functions requires an iterative method, such as iteratively reweighted least squares. In our simulations we standardized residuals using the median absolute deviation from the median. For details on this and other aspects of robust regression (e.g. confidence regions and tests for M-estimates) see Chapter 7 in Huber (1981). Applications of robust regression have been discussed, for instance, in Carroll (1980).

Notice that for small k, Huber's contrast function is very close to the objective function $\rho(x) = |x|$ used in L_1 regression. Among others, we will consider such a version of Huber's M-estimate and expect it to provide some insight concerning the performance of L_1-regression. We refer to Bassett Jr and Koenker (1978) for a more detailed discussion of L_1 regression.

In the next section of the paper we discuss the problem of model selection in the context of robust regression.

3.3 Robust model selection and the modified BIC

A natural way to obtain a robust version of the BIC (or the mBIC) is to replace the residual sum of squares in the criterion for model selection by a sum of contrasts. However, unlike in the L_2 case, a sum of contrasts will usually not be scale invariant. We therefore propose to standardize the Y_i in a robust way and

work with standardized observations, $Y_i^{(s)}$, obtained by subtracting the median and dividing by the median absolute deviation (MAD) as defined in Ronchetti et al. (1997). Notice that it is necessary to use the same estimate of the MAD in all the models considered, in order to make comparisons between models possible. We therefore propose to use the MAD calculated under the null model of no effect for the purposes of model selection. For the purpose of estimating the parameters of the various regression models, we additionally rescale the residuals separately for each model. In our simulations, we used the robust rescaling provided by the function 'rlm' in the library MASS of the R package, which is available under http://www.R-project.org.
This leads to the following robust version of the BIC:

$$BIC_\rho^* := n \log \sum_{i=1}^n \rho(Y_i^{(s)} - x_i'\hat{\theta}) + k \log(n) \ . \qquad (3.5)$$

Here, x_i' denotes the vector of regressors in the model (see 3.1) and $\hat{\theta}$ contains the regression coefficients estimated using $\rho(\cdot)$ as the contrast function.

An alternative approach proposed by Ronchetti et al. (1997) in the context of model selection is to rescale the contrast function instead of standardizing the Y_i's. As before, the same rescaling factors have to be used in all models. Ronchetti et al. (1997) propose estimation of the rescaling factors based on the largest possible model. This is not possible in our setup, since the largest possible model usually contains many more variables than observations. We therefore modified their approach and estimated rescaling constants under the null model. This way of rescaling the contrast function led to results that were almost identical to those obtained after standardizing Y using the MAD from the null model.

It has been shown by Machado (1993) that the robust BIC (3.5) is still consistent under quite general conditions on the error distribution. Martin (1980), as well as Ronchetti (1985), used similar ideas, in order to make the Akaike information criterion (AIC) robust. However, consistency is a minimal requirement, since the actual performance of BIC_ρ^* will depend both on $\rho(x)$ and the error distribution. Indeed this dependence becomes apparent from results in (Jurečková and Sen, 1996, see p. 410), who derived the limiting distribution of

$$\sum_{i=1}^{n} \left(\rho(Y_i^{(s)} - x_i'\hat{\theta}_1) - \rho(Y_i^{(s)} - x_i'\hat{\theta}_2) \right)$$

for a fixed null model M_1 versus a higher dimensional model M_2.

In order to obtain more reliable performance of the robust BIC for different contrast functions $\rho(x)$ and error distributions, it seems natural to renormalize BIC_ρ^* by taking the limiting distribution of

$$D_n = n(\log \sum \rho(Y_i^{(s)} - x_i'\hat{\theta}_1) - \log \sum \rho(Y_i^{(s)} - x_i'\hat{\theta}_2))$$

into account. Ideally, $BIC(M_2) - BIC(M_1)$ should have the same asymptotic distribution for different ρ and error models. For this purpose, we will derive the asymptotic distribution of D_n using results from Jurečková and Sen (1996), as well as the delta method. Since the asymptotic distribution of D_n depends not only on ρ, but also on the unknown error distribution, the required normalization constant needs to be estimated.

We compare model M_1 with parameter vector θ_1 of dimension $p_1 + q_1$ and model M_2 with parameter vector θ_2 of dimension $p_2 + q_2$. M_1 is assumed to be a submodel of M_2.

Theorem 1 *Let ρ be a contrast function satisfying the regularity conditions* specified in Chapter 5.5 of Jurečková and Sen (1996). Define $\tilde{Y}_i^{(s)}$ to be the observations standardized according to the population median and the population MAD. Furthermore, define the score function $\psi(x) = \rho'(x)$. Let $\gamma = \int \psi'(x)f(x)dx$, $\sigma_\psi^2 = \int \psi(x)^2 f(x)dx$ and $\delta = \int \rho(x)f(x)dx$, where $f(x)$ denotes the density function of the error distribution under the true regression model based on the observations $\tilde{Y}_i^{(s)}$. Moreover, let us define the constant $c_e = 2\gamma\delta/\sigma_\psi^2$.

Then under model M_1

$$c_e D_n \xrightarrow{d} \chi^2_{(p_2+q_2)-(p_1+q_1)}, \qquad (3.6)$$

as the sample size $n \to \infty$.

Proof. Let $l_n(\theta) = \sum \rho(Y_i^{(s)} - x_i'\theta)$, where θ is the true parameter vector. Note that

$$\frac{1}{n}l_n(\theta) = \frac{1}{n}\sum \rho(Y_i^{(s)} - x_i'\theta) \xrightarrow{p} E[\rho(\tilde{Y}_i^{(s)} - x_i'\theta)] = \delta.$$

From the consistency of M-estimates (see e.g. Huber (1981)) and the uniform continuity of l_n in a neighborhood of θ, it also holds that both

$$\frac{1}{n}l_n(\hat{\theta}_1) \xrightarrow{p} \delta, \quad \text{and} \quad \frac{1}{n}l_n(\hat{\theta}_2) \xrightarrow{p} \delta, \qquad (3.7)$$

where $\hat{\theta}_1$ and $\hat{\theta}_2$ are the M-estimates under models M_1 and M_2, respectively. Approximating D_n by the first term of its Taylor series expansion leads to

$$n(\log l_n(\hat{\theta}_1) - \log l_n(\hat{\theta}_2)) = n(\log(1 + \frac{l_n(\hat{\theta}_1) - l_n(\hat{\theta}_2)}{l_n(\hat{\theta}_2)})) \qquad (3.8)$$

$$= \frac{l_n(\hat{\theta}_1) - l_n(\hat{\theta}_2)}{\frac{1}{n}l_n(\hat{\theta}_2)} + R_n, \qquad (3.9)$$

where
$$R_n = O\left(\frac{(l_n(\hat{\theta}_1) - l_n(\hat{\theta}_2))^2}{\frac{1}{n}l_n^2(\hat{\theta}_2)}\right).$$

Jurečková and Sen (1996) (page 408-416, note their discussion of the extension of their results to studentized observations) proved that

$$\frac{2\gamma}{\sigma_\psi^2}(l_n(\hat{\theta}_1) - l_n(\hat{\theta}_2)) \xrightarrow{D} \chi^2_{(p_2+q_2)-(p_1+q_1)}. \tag{3.10}$$

Thus to conclude (3.6), it is enough to observe that from (3.7) and (3.10)

$$R_n \xrightarrow{p} 0$$

as $n \to \infty$.

□

For least squares regression, $\rho(Y_i^{(s)} - x_i'\theta)$ is equal to the residual sum of squares. In this situation, asymptotically D_n has a χ^2-distribution with $(p_2+q_2)-(p_1+q_1)$ degrees of freedom Serfling (1980). Hence, the normalization constant c_e is equal to 1 in this case.

As can be seen from Theorem 1, specific normalization constants (c_e) depend on the error distribution and the ρ-function considered. If the error distribution is assumed to be known, c_e can be derived analytically. The appropriate values for chosen distributions is shown in Table 3.1. In practice, the error distribution and c_e have to be estimated. For this purpose, we first carry out model selection with c_e equal to 1, i.e. the normalizing constant for Gaussian errors. The empirical distribution of the resulting residuals is then used to approximate the expected values, defining γ, σ_ψ^2 and δ by the corresponding averages (see e.g. page 409 Jurečková and Sen (1996)). Plugging in these quantities leads to the estimate $\hat{c}_e$. The discussion above leads us finally to the following robust version of the mBIC:

$$mBIC = \hat{c}_e n \log \sum_i \rho(Y_i^{(s)} - x_i'\hat{\theta}) + (p+q)\log n + \qquad (3.11)$$
$$2p \log(l-1) + 2q \log(u-1) ,$$

where ρ is a given contrast function and $\hat{\theta}$ is the corresponding M-estimate of the $(p+q)$-dimensional parameter vector of the model considered.

3.4 Comparison of performance under different error models

3.4.1 Design of the simulations

Simulations are carried out to compare the performance of least squares regression and robust methods for QTL mapping under a variety of error distributions. We consider M-estimates for robust models based on the following contrast functions: ρ_{Huber}, $\rho_{Bisquare}$ and ρ_{Hampel}. The parameters in the ρ-functions are set to $a = 2$, $b = 4$ and $c = 8$ for Hampel's function, $k = 1.345$ for Huber's function, and $c = 4.685$ for Tukey's bisquare M-estimator. These are the default parameter values used in the R-package MASS, which was used to obtain robust regression estimates. We also chose $k = 0.05$ for Huber's M-estimator as a close, smooth approximation to the L_1 contrast function $\rho(x) = |x|$. Notice that due to the smoothness of ρ_{Huber}, Theorem 1 still applies. The model selection process was carried out using the standard version of the mBIC (3.11) with $l = n_m/2.2$ and $u = n_e/2.2$. To solve the problem of searching over a large class of possible models, we use forward selection.

Three arrangements of marker genotypes with a backcross population of 200 individuals were simulated.

Arrangement 1: One chromosome of length 100 cM with 5 equally spaced markers;

Arrangement 2: Two chromosomes of length 100 cM both with 11 equally spaced markers;

Arrangement 3: Five chromosomes of length 100 cM each with 11 equally spaced markers.

Two scenarios were considered for each arrangement: A null model with no effects and a "3 QTL" model with one main effect of size $\beta = 0.55$ and one interaction effect (involving two loci) of size $\gamma = 1.2$. We assumed all the QTL to be located at marker positions.

All the methods were applied to each of the arrangements under each of six different error distributions. 1000 replications were used for arrangements 1 and 2, whereas we carried out 500 simulations for arrangement 3, which is computationally quite demanding. The performance of each method was measured by the average number of correctly identified main and epistasis effects, as well as the false discovery rate defined as

$$FDR = \frac{1}{n} \sum_{i=1}^{n} \frac{fp.m_i + fp.e_i}{fp.m_i + fp.e_i + c.m_i + c.e_i},$$

where the quantities $fp.m_i$ and $fp.e_i$ denote the number of false positive main and epistasis effects that were detected in replication i, and $c.m_i$ and $c.e_i$ are the number of correctly identified main and epistasis effects, respectively. According to the definition of the FDR (see Benjamini and Hochberg (1995)), the terms in the sum corresponding to replicates with no detections are set to be equal to zero. Under the null model, the false discovery rate is equivalent to the multiple type I (or familywise) error of detecting at least one incorrect effect. An inferred main effect was classified as being a false positive, if it was more than 15 cM away from the true QTL or the QTL had already been detected. An epistatic effect was classified as being a false positive, if at least one of the two QTL involved was more than 15 cM from the true QTL. Notice that this definition is fairly strict, since effects that are not very strong often lead to the detection of markers that are further than 15 cM away from the true QTL (see Bogdan and Doerge (2005) for a discussion).

3.4.2 Error distributions

We considered the following error distributions, which were all centered around the origin and standardized such that the inter-quartile range $(IQR = Q_{75} - Q_{25})$ was 1.5.

1. Normal: $1.11 \times N(0, \sigma^2)$ with $\sigma = 1$

2. Laplace (double exponential): $1.08 \times \frac{1}{2\lambda} exp(-|\lambda t|)$ with $\lambda = 1$

3. $Cauchy(scale = 0.75)$ with $scale = 0.5 \times IQR$

4. Tukey's gross error model: $1.081 \times (\lambda N(0, \sigma^2) + (1-\lambda) N(0, \tau\sigma^2))$ with $\lambda \sim Bernoulli(p = 0.95)$, $\sigma = 1$ and $\tau = 100$

5. χ^2 centered around the mean with 6 d.f.: $0.342 \times (\chi_6^2 - 6)$

6. χ^2 centered around the median with 6 d.f.: $0.342 \times (\chi_6^2 - \tilde{x}_6)$ with $\tilde{x}_6 = 5.348$

3.4.3 Results of the simulations and discussion

We focus on two points in particular. The first issue is to investigate whether our approach of using estimated normalization constants leads to similar values

Table 3.1: Values for normalization constants

error distr.	Huber$_{k=0.05}$	Huber$_{k=1.345}$	Bisquare	Hampel
Normal	1.267	1.079	1.095	1.025
Laplace	1.967	1.397	1.387	1.254
Cauchy	*	*	2.242	2.428
Tukey	1.770	1.952	1.408	1.653
χ^2	1.199	1.153	1.164	1.125
χ^2_{med}	1.295	1.255	1.248	1.165

* In the case of the Cauchy distribution, the integral for ρ_{Huber} leading to δ is infinite. The definitions of the error distributions are given in Section 3.4.2.

for the multiple type I error and the false discovery rate under various models for the error. We call this property error–robustness. The second issue is whether the power of our procedure for model selection remains high under non-normal errors and when outliers are present. We call this property power–robustness.

Table 3.2: Multiple type I errors for Arrangement 1.

	error distributions					
estimate	Normal	Laplace	Cauchy	Tukey	Chisq	Chisq-med
Huber$_{k=0.05}$ theor.	10.3	11.6	*	11.7	7.3	9.6
Huber$_{k=0.05}$ est.	10.2	12.0	9.7	10.3	7.6	10.5
Huber$_{k=1.34}$ theor.	11.1	10.1	*	15.5	10.7	13.0
Huber$_{k=1.34}$ est.	12.8	12.1	10.6	10.9	10.6	11.9
Bisquare theor.	11.1	10.6	12.0	13.6	10.5	12.3
Bisquare est.	12.2	12.2	11.4	12.0	10.4	10.6
Hampel theor.	10.8	11.1	12.4	14.6	12.3	12.8
Hampel est.	11.9	11.3	9.5	12.0	11.0	12.3
L_2^{mBIC}	12.9	10.6	4.6	7.3	11.1	12.3
L_2^{BIC}	26.4	22.1	14.9	18.9	23.5	24.0

Comparison of the probability of type I errors under the null model when the distribution of residuals is assumed to be known (theor., c_e) or has to be estimated (est., $\hat{c}_e$). For L_2 regression, c_e is always equal to 1. Definitions of the error distributions are given in Section 3.4.2.

We start by investigating the error–robustness. The probability of type I errors under the rules considered for selecting a model can be found in Tables 3.2–3.4. The two rows associated with each of the procedures present the results in the

Table 3.3: Multiple Type I errors for Arrangement 2.

estimate	\multicolumn{6}{c}{error distributions}					
	Normal	Laplace	Cauchy	Tukey	Chisq	Chisq-med
$\text{Huber}_{k=0.05}$ theor.	3.4	6.8	*	3.5	2.2	4.3
$\text{Huber}_{k=0.05}$ est.	4.2	5.3	6.2	4.6	5.6	5.6
$\text{Huber}_{k=1.34}$ theor.	4.4	5.1	*	7.2	4.7	7.1
$\text{Huber}_{k=1.34}$ est.	3.6	5.5	4.5	5.8	6.0	6.1
Bisquare theor.	4.1	4.1	4.6	4.8	4.5	6.7
Bisquare est.	3.4	5.0	5.4	5.2	5.4	5.8
Hampel theor.	4.2	5.6	5.9	6.3	5.1	6.5
Hampel est.	4.2	4.6	4.7	5.0	6.5	6.7
L_2^{mBIC}	4.4	4.6	8.8	3.4	6.1	5.7
L_2^{BIC}	90.4	89.8	81.6	86.9	88.4	88.6

Comparison of the probability of type I errors under the null model when the distribution of residuals is assumed to be known (theor., c_e) or has to be estimated (est., $\hat{c}_e$). For L_2 regression, c_e is always equal to 1. Definitions of the error distributions are given in Section 3.4.2.

cases when the proper theoretical constant was used and when the constant was estimated (separately for each replicate), respectively. These tables show that the multiple type I errors using the mBIC and estimated normalizing constants are comparable for all of the procedures. As expected from the formulae given in Bogdan et al. (2004), the probability of multiple type I errors for Arrangement 1 (only 5 markers) are approximately 0.1 and lower for Arrangements 2 and 3. For most of the examples there is only a slight difference between the results obtained using theoretical and estimated constants. The slightly larger difference obtained for the procedure using the Huber contrast under both the Laplace and heavy tailed distributions results from the problem of estimating the density of the Laplace distribution close to 0 (the Huber contrast with $k = 0.05$ assigns a relatively large weight to the corresponding residuals) and the tails of the Tukey distribution (the Huber contrast function tends to infinity as $x \to \pm\infty$). Also, note that the multiple type I error of the procedure for model selection based on the original BIC is much larger than desired and rapidly

Table 3.4: Multiple Type I errors for Arrangement 3.

estimate	Normal	Laplace	Cauchy	Tukey	Chisq	Chisq-med
			error distributions			
Huber$_{k=0.05}$ theor.	2.6	5.6	*	3.8	1.4	2.6
Huber$_{k=0.05}$ est.	5.6	6.8	6.2	3.4	6.4	6.0
Huber$_{k=1.34}$ theor.	6.4	5.2	*	9.7	2.0	6.6
Huber$_{k=1.34}$ est.	6.0	5.1	4.6	4.3	4.0	5.0
Bisquare theor.	3.8	4.4	3.6	4.0	3.4	5.0
Bisquare est.	4.8	2.6	3.8	4.6	5.0	5.0
Hampel theor.	5.0	7.2	5.6	7.2	4.4	6.2
Hampel est.	6.4	5.7	4.8	4.8	5.2	6.2
L_2^{mBIC}	5.5	2.0	6.5	3.0	3.5	3.5
L_2^{BIC}	100	100	99.0	100	100	100

Comparison of the probability of type I errors under the null model when the distribution of residuals is assumed to be known (theor., c_e) or has to be estimated (est., $\hat{c}_e$). For L_2 regression, c_e is always equal to 1. Definitions of the error distributions are given in Section 3.4.2.

increases as the number of regressors becomes larger. This demonstrates the advantage of using the mBIC rather than the BIC.

The false discovery rates under the "3 QTL" model can be found in Figures 3.1 to 3.3. They provide a similar picture, showing that our estimated normalizing constants lead to similar false discovery rates for several models for the error term. Indeed, the false discovery rates when applying the mBIC are approximately 15% for Arrangement 1 and 10% for Arrangements 2 and 3 (recall that a proportion of these "false" positives are due to the problem of localizing a QTL accurately). Least squares regression, in combination with standard BIC, results in false discovery rates above 20% for Arrangement 1, around 60% for Arrangement 2 and close to 100% for Arrangement 3, which again demonstrates the necessity of modifying the original BIC.

Figures 3.1 to 3.3 also provide information regarding the power–robustness. They present the average detection power (averaged over additive and interaction effects) and the estimated false discovery rate for the analyzed

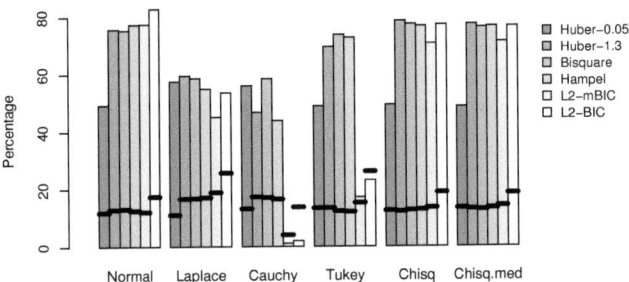

Figure 3.1: Percentage of correctly identified main and epistatic effects (shaded bars) and false discovery rates (horizontal black lines) for Arrangement 1. Definitions of the error distributions are given in Section 3.4.2

procedures under different error distributions. In the case of normal errors, model selection based on least squares regression and M-estimation performs comparably for all three arrangements. The only robust regression method which is significantly worse than L_2 regression under normal errors is the one based on the Huber contrast with $k = 0.05$. This confirms the low efficiency of L_1 regression under normality. The Huber contrast function with $k = 0.05$ also performed significantly worse than the other robust methods in the case of the Tukey and χ^2 error distributions. Our simulations demonstrate that standard L_2 regression performs relatively well under the Laplace and χ^2 distributions, while it is inferior to some of the robust methods. The largest difference between L_2 regression and robust methods is observed for the heavy-tailed Cauchy distribution (for which L_2 regression fails completely) and the Tukey distribution, according to which an outlier occurs with some given probability. Tukey's bisquare estimate performed well in all the problems considered.

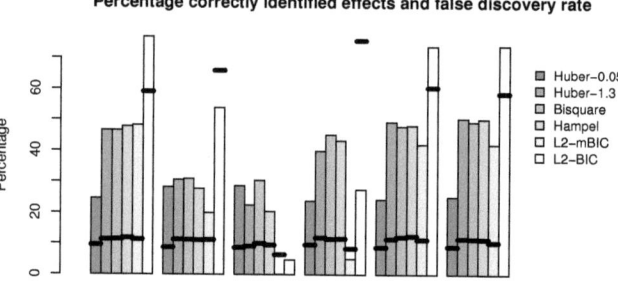

Figure 3.2: Percentage of correctly identified main and epistatic effects (shaded bars) and false discovery rates (horizontal black lines) for Arrangement 2. Definitions of the error distributions are given in Section 3.4.2

3.5 Application to real data

We apply our method to a data set obtained from QTL experiments on mice. Mähler et al. (2002) analyzed the susceptibility to colitis in strains that carry a deficient IL-10 gene, which is important in controlling the response of the immune system to intestinal antigenes. We consider their data obtained from a backcross to the less susceptible B6 strain and the quantitative traits MidPC1 and CecumPC1, which are the first two principal components of four scores measuring the severity and type of lesions in middle colon and cecum, respectively.

The data set contains 203 individuals and twelve markers from nine chromosomes, which were selected from a preliminary genome scan of 40 individuals and 67 markers spread over all 20 chromosomes.

Mähler et al. (2002) found one significant main effect for the MidPC1 trait on chromosome 12, which explains 6.7% of the variance and one possible main effect

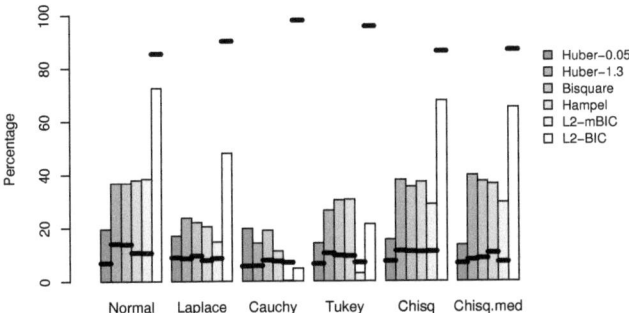

Figure 3.3: Percentage of correctly identified main and epistatic effects (shaded bars) and false discovery rates (horizontal black lines) for Arrangement 3. Definitions of the error distributions are given in Section 3.4.2

for CecumPC1 on chromosome 13, which explains 3.4% of the variance, but no epistatic effects. The distributions of the residuals under the selected model clearly deviated from normality by being bimodal in the case of both traits. The modified BIC based on both least squares and M-estimation confirmed the main effect for ModPC1. In addition, both methods found an epistatic effect between the marker on chromosome 4 at 71 cM and the marker on chromosome 7 at 46 cM. In the case of CecumPC1, model selection using least squares regression found no QTL. Using M-estimation with a bi-square, Huber or Hampel contrast function, we detected two QTL, one on each of chromosome 5 and chromosome 13. The effect on chromosome 5 is slightly stronger and has a different sign to the effect on chromosome 13, as suggested by Mähler et al. (2002).

This example illustrates that our robust methods of model selection are capable of finding additional effects when the distribution of errors is not normal.

3.6 Conclusions

Overall, the performance of the M-estimators considered is superior to least squares regression in the context of QTL mapping under various conditions. Considering the wide spectrum of possible error distributions, M-estimates based on the contrast functions considered also prove to be more flexible than L_1 regression. Among the robust methods considered, Tukey's bisquare estimate showed particularly good overall performance.

Chapter 4

Locating multiple interacting quantitative trait loci using rank-based model selection

Based on:

Zak, M., Baierl, A., Bogdan, M. and Futschik, A. (2007). Locating multiple interacting quantitative trait loci using rank-based model selection. *Genetics* **176**, 1845–1854

4.1 Introduction

An alternative solution to the problem of non-normality of the error distribution is provided by nonparametric methods based on ranks. In the context of QTL mapping, this approach has been proposed and investigated e.g. in Kruglyak and Lander (1995), Broman (2003), or Zou et al. (2003). A major advantage of rank based statistics is that their distribution under the "null" hypothesis does not depend on the error distribution. Moreover, as demonstrated in Zou et al. (2003) (see also Lehmann (1975)), the asymptotic efficiency of rank tests is only slightly smaller than that of the classical tests when the error distribution is normal, and much higher when the the error distribution is heavy tailed.

The advantage of rank-based methods over M-estimation discussed in Chapter 3 lies in the smaller computational effort of least squares regression. This can become relevant in the context of QTL mapping, where the verification of a large number of competing models is required.

In this chapter, we use the idea of rank tests and propose a new version of the mBIC which is based on ranks instead of the original trait values. For continuous error distributions and for the standard null model of no effects, we prove that the asymptotic distribution of the rank version of the mBIC is the same as the null distribution of the regular mBIC for normal errors.

4.2 Methods

When applying rank based model selection, one exchanges the trait values by their ranks. A major advantage of using ranks is that the distribution of the test statistic under the null hypothesis of no QTL does not depend on the distribution of the error terms. Using ranks strongly reduces the influence of heavy tails and outlying observations.

Our proposed construction of rBIC, the rank version of the mBIC, is very simple. After substituting the trait values by their ranks, we calculate the rank residual sum of squares, $rRSS = \sum_{i=1}^{n}(R_i - \tilde{R}_i)^2$, where $\tilde{R} = (X'X)^{-1}X'R$ and $X = (\mathbf{1}, X_1, \ldots, X_n)'$, with $\mathbf{1} = (1, \ldots, 1)'$. By replacing the RSS term with $rRSS$ in formula (2.3), we obtain

$$rBIC = n\log(rRSS) + (p+r)\log(n) + 2p\log(l-1) + 2r\log(u-1), \quad (4.1)$$

with all the parameters denoted as before.

4.2.1 Simulation design

We consider two setups with QTL and marker genotypes of backcross populations of size 200 in case of setup 1 and 200 and 500 for setup 2. Setup 1 is identical with Arrangement 2 described in Section 3.4.1 in order to allow a comparison of robust and rank-based model selection, respectively.

In the second setup, we consider three chromosomes each of length 100 cM with 7, 8 and 7 markers, respectively, distributed randomly across the chromosome. The distances between the markers range from 1 to 29 cM with a mean distance of 15.79 cM. To narrow these intervals and to enable a location of QTL at a finer scale, we used regression interval mapping according to Haley and Knott (1992). This method relies on imputing putative QTL between markers and to replace their missing genotypes by expected values, calculated on the basis of neighboring markers. Using this approach, we imputed additional marker genotypes in order to reduce the intervals between adjacent markers to a maximum of 10 cM. The second setup is considered under the null model of no effects and an alternative model involving 3 main and 3 epistatic effects. The locations and the sizes of main QTL effects are as follows: QTL1 on chromosome 1 at 20 cM with $\beta_1 = 0.8$, QTL2 on chromosome 2 at 20 cM with $\beta_2 = 0.7$ and QTL3 on chromosome 3 at 1 cM with $\beta_3 = 0.6$. The epistatic effects are specified as follows: interaction 1 involving QTL1 and QTL3 with $\gamma_1 = 1.6$, interaction 2 involving QTL2 and a new QTL on chromosome 3 at 75 cM with $\gamma_2 = 1.4$ and interaction 3 involving two new QTL, both on chromosome 1 at 27 and 60 cM, respectively, with $\gamma_3 = 1.2$.

The simulations were performed using the mBIC and rBIC criterion. We apply the standard forms of these criteria with $l = N_m/2.2$ and $u = N_e/2.2$. To solve the problem of searching over a large class of possible models, we use forward selection. The simulation results are based on 3000 replications.

To investigate the robustness of our proposed criterion, we consider noise distributions 1 to 5 defined in Section 3.4.2.

In Tables 4.3 and 4.4, the average number of correctly identified terms and the false discovery rate (FDR, see Section 3.4.1 for definition) are presented. In Table 4.3, a main effect is assumed to be correctly identified if at least one of the chosen markers is within 15 cM of the true QTL. Every additional marker within this range that is selected is counted as false positive. An epistatic effect is assumed to be correctly identified if both markers of the chosen interaction term are within 15 cM of the respective QTL. In Table 4.4, the threshold for correct identification is increased to 30 cM. This large detection window is particularly suitable for $n = 200$ since in this case the standard deviation of the localization of QTL with magnitudes according to our simulated effect sizes is close to 10cM. In case of the Cauchy distribution it reaches even 15 cM. These estimates were obtained by additional simulations.

4.2.2 Simulation results

For setup 1, The type I errors under the null model of no effects are summarized in Table 4.1. The differences between the results for the mBIC and the rBIC depend on the noise distribution and are small in most cases (Cauchy noise is an exception). According to Proposition 1, the slightly different values obtained for different error distributions with the rBIC are due to random simulation errors. In the context of Setup 1, we compare the power and FDRs of our proposed rank based method to the M-estimates investigated by Baierl et al. (2007) as well as to the classical BIC. In regression, M-estimates are obtained by minimizing more general measures of distance instead of the residual sum of squares. In Baierl et al. (2007), the following three contrast functions have been considered as a measure of distance: Huber's, Bisquare and Hampel's contrast function. The results of their simulations indicate that the use of the above mentioned robust contrast functions leads to much better results than those obtained by least squares regression in cases when the error terms come from a heavy-tailed distribution. In the normal case, both methods work comparably. The results for the first setup in the case of two effects are presented in Figures 4.1 (average percentage of correctly identified effects) and 4.2 (FDR). The horizontal lines indicate the values obtained for the rank based method. In the case of non-normal distributions, the percentage of correctly identified effects is in most cases higher for the rank method than for the other methods. None of the M-estimators performs significantly better for every type of noise. In Figure 4.2 on the other hand, we observe that the rank method performs similar to M-estimators but leads to slightly higher FDRs. Overall, it is impossible to choose a robust method that will perform consistently better than the others for all noise distributions. However, the rank based rBIC seems to perform well in

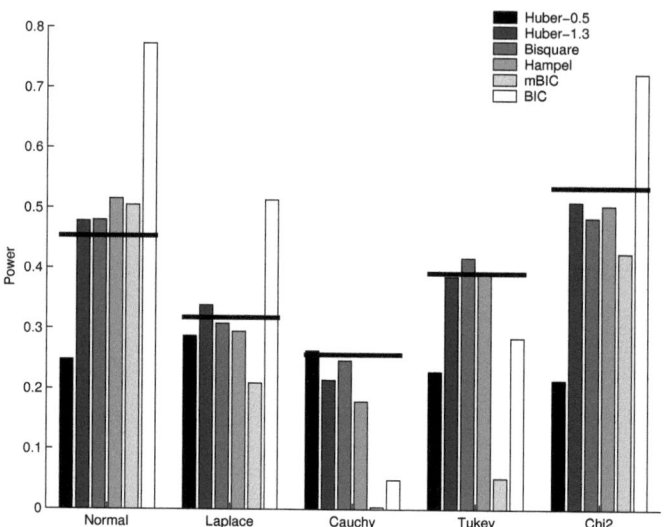

Figure 4.1: Percentage correctly identified main and epistatic effects for robust methods (shaded bars) and rank based method (horizontal black lines)

all settings. What's more, the method is very simple to use and computationally less demanding than M-estimates.

Notice that the original BIC criterion leads to a considerably higher percentage of correct identification but also (see Figure 4.2) to extremely high false discovery rates.

Next we will consider the second setup which is more realistic from a practical point of view. Our simulations indicate that the type I error is smaller in most

Table 4.1: Type I errors under the null model (no QTL) for setup 1.

criterion	error distribution				
	Normal	Laplace	Cauchy	Tukey	Chi2
mBIC	0.057	0.052	0.075	0.047	0.058
rBIC	0.055	0.055	0.054	0.057	0.062

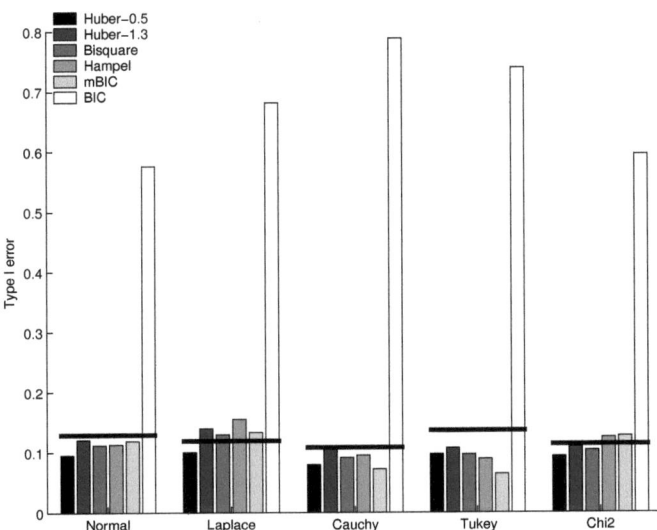

Figure 4.2: False discovery rates for robust (shaded bars) and rank based methods (horizontal black lines)

cases for the rBIC than for the mBIC (see Table 4.2). The largest differences are observed for the Cauchy and Tukey error distributions.

For the 6-effect model in setup 2 and a 15 cM identification window, the FDR for the rBIC ranges from 12% to 15% for $n = 200$ and from 3% to 8 % for $n = 500$ (see Table 4.3). The relatively large FDR values for $n = 200$ are caused by a large standard deviation of the estimates of QTL location. Our additional

Table 4.2: **Type I errors under the null model (no QTL) for setup 2.**

		error distribution				
n	criterion	Normal	Laplace	Cauchy	Tukey	Chi2
200	mBIC	0.031	0.029	0.085	0.046	0.032
200	rBIC	0.030	0.031	0.022	0.030	0.030
500	mBIC	0.015	0.021	0.079	0.028	0.024
500	rBIC	0.015	0.022	0.018	0.020	0.021

simulations demonstrated that this standard error reaches 15 cM for our simulated QTL and both sample sizes when the noise is Cauchy-distributed. Thus a significant proportion of "false positives" is due to correctly identified but imprecisely localized QTL. This is confirmed by our results provided in Table 4.4 where the size of the identification window is chosen to be 30 cM. Applying this more liberal identification criterion, the FDR for the rBIC is at a level of 3-8% for $n = 200$ and at a level of 0.5-1.5% for $n = 500$. Table 4.4 also demonstrates that for the Cauchy and Tukey error distribution the FDR for the rBIC is significantly smaller than the FDR for the standard mBIC. For other noise distributions, the FDR of the rBIC is comparable to the corresponding values for the mBIC.

We now compare the power of the rBIC to that of the mBIC in the context of 30 cM identification windows (see Table 4.4). The results demonstrate that the rBIC is slightly less efficient than the mBIC if the error distribution is normal. The corresponding loss of power is equal to 4 percentage points for $n = 200$ (from 45% to 41%) and to 2 percentage points for $n = 500$ (from 90% to 88%). For all other investigated error distributions, the rBIC has a larger power than the mBIC. A particularly large difference occurs for the Tukey distribution where for $n = 500$ the power of the rBIC is 82% compared to 18% for the mBIC. For the Cauchy distribution the mBIC completely fails (the power is below 1%) and the power of the rBIC for $n = 500$ is equal to 55%. Note that both the Tukey and Cauchy distribution, lead to a certain proportion of outliers. The results confirm that the rank based method works comparably well for normal errors and much better when outliers are present.

Table 4.3: **Results for setup 2 (6-effect-model) and a 15cM identification window**

	n = 200				n=500			
	mBIC		rBIC		mBIC		rBIC	
noise	FDR	%corr	FDR	%corr	FDR	%corr	FDR	%corr
1 N	0.117	0.408	0.123	0.369	0.033	0.876	0.034	0.858
2 L	0.131	0.185	0.142	0.237	0.070	0.625	0.055	0.721
3 C	0.091	0.006	0.148	0.135	0.077	0.004	0.082	0.511
4 T	0.101	0.054	0.136	0.292	0.111	0.161	0.043	0.790
5 $\chi^2 - 6$	0.129	0.350	0.121	0.382	0.037	0.841	0.034	0.864

Table 4.4: **Results for setup 2 (6-effect model) and a 30cM identification window**

	n = 200				n=500			
	mBIC		rBIC		mBIC		rBIC	
noise	FDR	%corr	FDR	%corr	FDR	%corr	FDR	%corr
1 N	0.029	0.452	0.035	0.409	0.004	0.902	0.005	0.883
2 L	0.041	0.211	0.042	0.269	0.013	0.663	0.008	0.756
3 C	0.081	0.010	0.046	0.162	0.072	0.007	0.015	0.549
4 T	0.054	0.064	0.034	0.329	0.036	0.182	0.009	0.818
5 $\chi^2 - 6$	0.032	0.390	0.030	0.422	0.006	0.869	0.005	0.890

4.2.3 Application to Real Data

The same data set by Mähler et al. (2002) as described in Section 3.5 was used to verify the performance of the rBIC in the case of real data.

Preliminary to applying our method to the data set, we removed 16 and 15 observations for the analysis of trait CecumPC1 and MidPC1, respectively, due to missing trait or genotype information. Further we excluded marker D17Mit88, which had missing genotypes for 62 individuals. Imputation of missing genotype data was not feasible because of the low marker density.

The considered traits are summaries of discrete measures (scores). For 187 observations of the CecumPC1 there are 32 different trait values, 45% of the observations fall within one of four most frequent values and the most numerous group contains 13% of the observations. There are also only 19 different values for the MidPC1. Among the 188 observations of this trait, 42% are equal to the most frequent value and 11% to the second frequent. In order to derive ranks for individuals with identical trait values, midranks as discussed in Section 3 were calculated.

Since we do not have any prior information we use the standard versions of the mBIC and the rBIC with $l = N_m/2.2$ and $u = N_e/2.2$.

Applying both the mBIC and the rBIC to the MidPC1 data set, we find two effects, one main and one epistatic. The main effect found by our approach is the same as in Mähler et al. (2002). However, we also detected an epistatic effect between markers on chromosomes 4 and 7 that considerably improves the fit of the model to the data. The fraction of the variance explained by the model, the R^2, increases from 0.0768 for the one effect model to 0.1397 for the model which also includes the interaction term.

For the second trait, CecumPC1, the mBIC does not find any effect. When using the rBIC on the other hand, we get one main effect on chromosome 5 (D5Mit205). This effect is different from the one that was suggested by Mähler et al. (2002). As can be seen in Figure 4.3, the value of the rank statistic for the effect found by Mähler et al. (2002) labeled with number 9 is substantially smaller than the one found by the rBIC, labeled with number 3. The respective p-values of the Wilcoxon test are 0.0089 and 0.0026, which supports the choice of marker number 3. For the t-test, the p-values are 0.0061 for marker 9 and 0.0066 for marker number 3. When correcting for multiple testing, none of these values is significant and none of these effects is detected by the regular mBIC criterion. The marker D5Mit205 was also detected by the robust version of mBIC (see Section 3.5), which additionally detects effect number 9.

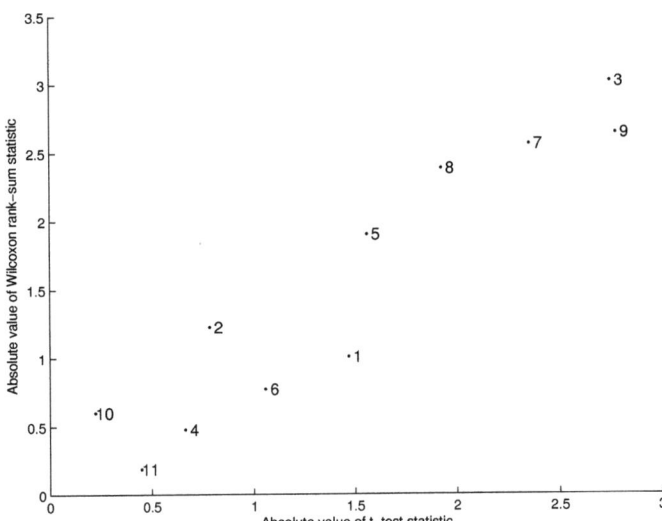

Figure 4.3: Absolute values of the t-test statistics vs. absolute values of the Wilcoxon statistics, for the 11 markers used in the analysis of the CecumPC1.

Bibliography

Akaike, H. (1973). Information theory and an extension of the maximum likelihood principle. *International Symposium on Information Theory, 2 nd, Tsahkadsor, Armenian SSR*, pages 267–281.

Baierl, A., Bogdan, M., Frommlet, F., and Futschik, A. (2006). On Locating Multiple Interacting Quantitative Trait Loci in Intercross Designs. *Genetics*, 173(3):1693–1703.

Baierl, A., Futschik, A., Bogdan, M., and Biecek, P. (2007). Locating multiple interacting quantitative trait loci using robust model selection. *Computational Statistics and Data Analysis*.

Ball, R. (2001). Bayesian Methods for Quantitative Trait Loci Mapping Based on Model Selection: Approximate Analysis Using the Bayesian Information Criterion. *Genetics*, 159(3):1351–1364.

Bassett Jr, G. and Koenker, R. (1978). Asymptotic Theory of Least Absolute Error Regression. *Journal of the American Statistical Association*, 73(363):618–622.

Benjamini, Y. and Hochberg, Y. (1995). Controlling the false discovery rate: a practical and powerful approach to multiple testing. *Journal of the Royal Statistical Society. Series B. Methodological*, 57(1):289–300.

Bogdan, M. and Doerge, R. (2005). Biased estimators of quantitative trait locus heritability and location in interval mapping. *Heredity*, 95:476–484.

Bogdan, M., Ghosh, J., and Doerge, R. (2004). Modifying the Schwarz Bayesian Information Criterion to Locate Multiple Interacting Quantitative Trait Loci. *Genetics*, 167(2):989–999.

Broman, K. (1997). *Identifying Quantitati e Trait Loci in Experimental Crosses*. PhD thesis, PhD thesis, Department of Statistics, University of California, Berkeley, CA, USA.

Broman, K. (2001). Review of statistical methods for QTL mapping in experimental crosses. *Lab Anim (NY)*, 30(7):44–52.

Broman, K. (2003). Mapping Quantitative Trait Loci in the Case of a Spike in the Phenotype Distribution. *Genetics*, 163(3):1169–1175.

Broman, K. and Speed, T. (2002). A model selection approach for the identification of quantitative trait loci in experimental crosses (with discussion).

Bulmer, M. (1980). *The mathematical theory of quantitative genetics*. Clarendon Press.

Carlborg, Ö. and Haley, C. (2004). Epistasis: too often neglected in complex trait studies. *Nat Rev Genet*, 5(8):618–625.

Carroll, R. (1980). A Robust Method for Testing Transformations to Achieve Approximate Normality. *Journal of the Royal Statistical Society. Series B (Methodological)*, 42(1):71–78.

Clyde, M. (1999). Bayesian model averaging and model search strategies. *Bayesian Statistics*, 6:157–185.

Davison, A. (2003). *Statistical models*. Cambridge University Press Cambridge.

Doerge, R. (2002). Mapping and analysis of quantitative trait loci in experimental populations. *Nat Rev Genet*, 3(1):43–52.

Dorigo, M., Caro, G., and Gambardella, L. (1999). Ant Algorithms for Discrete Optimization. *Artificial Life*, 5(2):137–172.

Fijneman, R. (1998). High frequency of interactions between lung cancer susceptibility genes in the mouse: mapping of Sluc5 to Sluc14. *Cancer Research*, 58(21):4794–4798.

Fijneman, R., De Vries, S., Jansen, R., and Demant, P. (1996). Involvement of quantitative trait loci in complex interactions: mapping of four new loci, Sluc1, Sluc2, Sluc3, and Sluc4, that influence susceptibility to lung cancer in the mouse. *Nature (Genet.)*, 14:465–467.

Fisher, R. (1918). The Correlation Between Relatives on the Supposition of Mendelian Inheritance. *Trans. Royal Soc. Edinburgh*, 52:399–433.

Galton, F. (1869). *Hereditary Genius: An Inquiry into Its Laws and Consequences*. Reprinted 1962, Meridian, New York.

Galton, F. (1889). *Natural Inheritance*. Macmillan London.

George, E. and McCulloch, R. (1993). Variable Selection Via Gibbs Sampling. *Journal of the American Statistical Association*, 88(423):881–889.

Glover, F. (1989a). Tabu Search-Part I. *ORSA Journal on Computing*, 1(3):190–206.

Glover, F. (1989b). Tabu search-Part II. *ORSA Journal on Computing*, 2(1):4–32.

Goldberg, D. (1989). *Genetic Algorithms in Search, Optimization and Machine Learning*. Addison-Wesley Longman Publishing Co., Inc. Boston, MA, USA.

Haley, C. and Knott, S. (1992). A simple regression method for mapping quantitative trait loci in line crosses using flanking markers. *Heredity*, 69(4):315–24.

Huber, P. (1981). Robust Statistics, J. *Willey & Sons, NY*.

Huttunen, S., Aspi, J., Hoikkala, A., and Schlötterer, C. (2004). QTL analysis of variation in male courtship song characters in Drosophila virilis. *Heredity*, 92:263–269.

Jansen, R. (1993). Interval Mapping of Multiple Quantitative Trait Loci. *Genetics*, 135(1):205–211.

Jansen, R. and Stam, P. (1994). High Resolution of Quantitative Traits Into Multiple Loci via Interval Mapping. *Genetics*, 136(4):1447–1455.

Jurečková, J. and Sen, P. (1996). *Robust Statistical Procedures: Asymptotics and Interrelations*. Wiley, New York.

Kao, C. and Zeng, Z. (2002). Modeling Epistasis of Quantitative Trait Loci Using Cockerham's Model. *Genetics*, 160(3):1243–1261.

Kao, C., Zeng, Z., and Teasdale, R. (1999). Multiple Interval Mapping for Quantitative Trait Loci. *Genetics*, 152(3):1203–1216.

Kirkpatrick, S., Gelatt, C., and Vecchi, M. (1983). Optimization by simulated annealing. *Science*, 220:671–680.

Kruglyak, L. and Lander, E. (1995). A Nonparametric Approach for Mapping Quantitative Trait Loci. *Genetics*, 139(3):1421–1428.

Lander, E. and Botstein, D. (1989). Mapping Mendelian Factors Underlying Quantitative Traits Using RFLP Linkage Maps. *Genetics*, 121(1):185–199.

Lehmann, E. (1975). *Nonparametrics: Statistical Methods Based on Ranks*. Holden-Day, Inc., San Francisco, Ca.

Lush, J. (1937). *Animal Breeding Plans*. Iowa State Univ. Pres, Ames.

Lynch, M. and Walsh, B. (1998). *Genetics and analysis of quantitative traits*. Sinauer Sunderland, Ma.

Machado, J. (1993). Robust model selection and M-estimation. *Econometric Theory*, 9(3):478–93.

Mähler, M., Most, C., Schmidtke, S., Sundberg, J., Li, R., Hedrich, H., and Churchill, G. (2002). Genetics of Colitis Susceptibility in IL-10-Deficient Mice: Backcross versus F2 Results Contrasted by Principal Component Analysis. *Genomics*, 80(3):274–282.

Martin, R. (1980). Robust estimation of autoregressive models. *Directions in Time Series*, pages 228–262.

Phillips, T. (2002). Animal Models for the Genetic Study of Human Alcohol Phenotypes. *Alcohol Research & Health*, 26(3):202–209.

Piepho, H. and Gauch, H. (2001). Marker Pair Selection for Mapping Quantitative Trait Loci. *Genetics*, 157(1):433–444.

Ping, Z. (1997). Comment to Shao, J.: An asymptotic theory for linear model selection. *Statistica Sinica*, 7(2):254–258.

Ronchetti, E. (1985). Robust model selection in regression. *STAT. PROB. LETT.*, 3(1):21–24.

Ronchetti, E., Field, C., and Blanchard, W. (1997). Robust Linear Model Selection by Cross Validation. *Journal of the American Statistical Association*, 92(439).

Schwarz, G. (1978). Estimating the Dimension of a Model. *The Annals of Statistics*, 6(2):461–464.

Sen, S. and Churchill, G. (2001). A Statistical Framework for Quantitative Trait Mapping. *Genetics*, 159(1):371–387.

Serfling, R. (1980). *Approximation theorems of mathematical statistics*. Wiley New York.

Shimomura, K., Low-Zeddies, S., King, D., Steeves, T., Whiteley, A., Kushla, J., Zemenides, P., Lin, A., Vitaterna, M., Churchill, G., et al. (2001). Genome-Wide Epistatic Interaction Analysis Reveals Complex Genetic Determinants of Circadian Behavior in Mice. *Genome Research*, 11(6):959–980.

Siegmund, D. (2004). Model selection in irregular problems: Applications to mapping quantitative trait loci. *Biometrika*, 91(4):785–800.

Thompson, E. (2000). Statistical Inferences from Genetic Data on Pedigrees. *NSF-CBMS Regional Conference Series in Probability and Statistics*, 6.

van de Ven, R. (2004). Reversible-Jump Markov Chain Monte Carlo for Quantitative Trait Loci Mapping.

Wolf, J., Brodie, E., and Wade, M. (2000). *Epistasis and the Evolutionary Process*. Oxford University Press US.

Wright, S. (1921a). Correlation and causation. *Journal of Agricultural Research*, 20(3):557–585.

Wright, S. (1921b). Systems of Mating. I. The Biometric Relations Between Parent and Offspring. *Genetics*, 6(2):111–123.

Wright, S. (1921c). Systems of Mating. II. The Effects of Inbreeding on the Genetic Composition of a Population. *Genetics*, 6(2):124–143.

Wright, S. (1921d). Systems of Mating. III. Assortative Mating Based on Somatic Resemblance. *Genetics*, 6(2):144–161.

Yi, N. and Xu, S. (2002). Mapping Quantitative Trait Loci with Epistatic Effects. *Genetical Research*, 79(02):185–198.

Yi, N., Xu, S., and Allison, D. (2003). Bayesian Model Choice and Search Strategies for Mapping Interacting Quantitative Trait Loci. *Genetics*, 165(2):867–883.

Yi, N., Yandell, B., Churchill, G., Allison, D., Eisen, E., and Pomp, D. (2005). Bayesian Model Selection for Genome-Wide Epistatic Quantitative Trait Loci Analysis. *Genetics*, 170(3):1333–1344.

Yohai, V. (1987). High Breakdown-Point and High Efficiency Robust Estimates for Regression. *The Annals of Statistics*, 15(2):642–656.

Zeng, Z. (1993). Theoretical Basis for Separation of Multiple Linked Gene Effects in Mapping Quantitative Trait Loci. *Proceedings of the National Academy of Sciences*, 90(23):10972–10976.

Zeng, Z. (1994). Precision Mapping of Quantitative Trait Loci. *Genetics*, 136(4):1457–1468.

Zou, F., Yandell, B., and Fine, J. (2003). Rank-Based Statistical Methodologies for Quantitative Trait Locus Mapping. *Genetics*, 165(3):1599–1605.

List of Figures

1.1 Types of experimental crosses. One pair of columns corresponds to the genome of one individual. The parents (F0–generation) have both identical pairs of chromosomes, maternal and paternal DNA is indicated by black and white bars, respectively. The children (F1) inherit one chromosome from their father and one from their mother. Backcross populations are achieved by crossing children with either father or mother. In intercross populations, grandchildren (F2) are produced. 14

1.2 Overview of available techniques for QTL mapping 17

2.1 Comparison of the Bonferroni type I error bounds under the null model (no effects) for the intercross design when the same penalty as in backcross is used and when the penalty is adjusted accordingly 27

2.2 The dark curves show the percentage of correctly identified additive, dominance and epistatic effects depending on the heritability. The grey shaded curves display the expected number of incorrectly selected (linked and unlinked) markers ($n = 200$). 29

2.3 Percentage of correctly identified additive effects vs. number of additive effects. The QTL are unlinked, i.e. located on different chromosomes, and have effect sizes of 0.5. The solid line is based on simulations where no prior information is used to derive the penalty terms of the modified BIC. The dashed line represents simulations with the correct number of underlying effects (1,2,4,7,10) assumed known. The dotted line corresponds to the two step search procedure based on Formula (2.4).................. 35

2.4 Percentage correctly identified additive, dominance and epistatic effects vs. individual effect heritabilities $h^2 = \sigma^2_{eff}/(\sigma^2_\epsilon + \sigma^2_G)$. Detection rates are taken from simulations of scenarios 1-9 (see Table 2.3) for $n = 200$. 39

2.5 Percentage correctly identified additive, dominance and epistatic effects vs. individual effect heritabilities $h^2 = \sigma^2_{eff}/(\sigma^2_\epsilon + \sigma^2_G)$. Detection rates are taken from simulations of scenarios 1-9. (see Table 2.3) for $n = 500$. 40

2.6 Genetic map for the *Drosophila v.* experiment by Huttunen et al. (2004). Solid horizontal lines indicate observed marker positions, dotted lines show imputed positions. QTL localized by our proposed method are symbolized by diamonds. Intervals with significant additive and/or dominance effects found by Huttunen et al. (2004) applying composite interval mapping are indicated by solid vertical lines. 42

3.1 Percentage of correctly identified main and epistatic effects (shaded bars) and false discovery rates (horizontal black lines) for Arrangement 1. Definitions of the error distributions are given in Section 3.4.2 63

3.2 Percentage of correctly identified main and epistatic effects (shaded bars) and false discovery rates (horizontal black lines) for Arrangement 2. Definitions of the error distributions are given in Section 3.4.2 64

3.3 Percentage of correctly identified main and epistatic effects (shaded bars) and false discovery rates (horizontal black lines) for Arrangement 3. Definitions of the error distributions are given in Section 3.4.2 65

4.1 Percentage correctly identified main and epistatic effects for robust methods (shaded bars) and rank based method (horizontal black lines) 72

4.2 False discovery rates for robust (shaded bars) and rank based methods (horizontal black lines) 73

4.3 Absolute values of the t-test statistics vs. absolute values of the Wilcoxon statistics, for the 11 markers used in the analysis of the CecumPC1. 77

List of Tables

2.1	Description of scenarios 1-9	36
2.2	Simulation results for sample size 200 and 500 (initial penalty)	37
2.3	Simulation results for sample size 200 and 500 (adjusted penalty)	38
2.4	Simulation results for different percentages of missing data	41
3.1	Values for normalization constants	60
3.2	Multiple type I errors for Arrangement 1.	60
3.3	Multiple Type I errors for Arrangement 2.	61
3.4	Multiple Type I errors for Arrangement 3.	62
4.1	Type I errors under the null model (no QTL) for setup 1.	72
4.2	Type I errors under the null model (no QTL) for setup 2.	73
4.3	Results for setup 2 (6-effect-model) and a 15cM identification window	75
4.4	Results for setup 2 (6-effect model) and a 30cM identification window	75

Südwestdeutscher Verlag für Hochschulschriften

Wissenschaftlicher Buchverlag bietet
kostenfreie
Publikation
von
Dissertationen und Habilitationen

Sie verfügen über eine wissenschaftliche Abschlußarbeit zu aktuellen oder zeitlosen Fragestellungen, die hohen inhaltlichen und formalen Anspruchen genügt, und haben **Interesse an einer honorarvergüteten Publikation?**

Dann senden Sie bitte erste Informationen über Ihre Arbeit per Email an: info@svh-verlag.de.

Unser Außenlektorat meldet sich umgehend bei Ihnen.

Südwestdeutscher Verlag für Hochschulschriften
Aktiengesellschaft & Co. KG
Dudweiler Landstr. 99
D – 66123 Saarbrücken
www.svh-verlag.de

Printed by Books on Demand GmbH, Norderstedt / Germany